AF890210

Maria Eletta Negretti

Hydrodynamic instabilities and entrainment in two-layer stratified flows down a slope

Dissertationsreihe am Institut für Hydromechanik
der Universität Karlsruhe (TH)
Heft 2007/1

Hydrodynamic instabilities and entrainment in two-layer stratified flows down a slope

von
Maria Eletta Negretti

universitätsverlag karlsruhe

Dissertation, genehmigt von der
Fakultät für Bauingenieur-, Geo- und Umweltwissenschaften
der Universität Fridericiana zu Karlsruhe (TH), 2007
Referenten: Prof. Gerhard H. Jirka, Ph.D., Prof. David Z. Zhu, Ph.D.,
Prof. Scott A. Socolofsky, Ph.D., Prof. Dr. Wolfgang Rodi,
Priv. Doz. Dr. Andreas Class, Prof. Dr. Jochen Fröhlich

Impressum

Universitätsverlag Karlsruhe
c/o Universitätsbibliothek
Straße am Forum 2
D-76131 Karlsruhe
www.uvka.de

Universitätsverlag Karlsruhe 2007
Print on Demand

ISSN: 1439-4111
ISBN: 978-3-86644-170-5

Hydrodynamic instabilities and entrainment processes in two-layer density-stratified exchange flows over a submerged sill

Zur Erlangung des akademischen Grades eines

DOKTOR-INGENIEURS

an der Fakultät für Bauingenieurwesen
der Universität Fridericiana zu Karlsruhe (TH)

eingereichte

DISSERTATION

von

Maria Eletta Negretti

aus Italien

Tag der mündlichen Prüfung: 11.07.2007

Hauptreferent:	Prof. Gerhard H. Jirka, Ph.D.
Koreferent:	Prof. David Z. Zhu, Ph.D.
	Prof. Scott A. Socolofsky, Ph.D.
	Prof. Dr. Wolfgang Rodi
	Priv. Doz. Dr. Andreas Class
	Prof. Dr. Jochen Fröhlich

Karlsruhe, 11. Juli 2007

A mia madre

Abstract

Fluid mechanical problems related to two-layer flows are of considerable interest in a variety of geophysical, atmospheric and engineering contexts. Experience has shown that the understanding of such problems can also yield valuable insight in stratified layered flow problems including power plant and reservoir outflows as well as harbor and flood channel design. While many geophysical and atmospheric flows involve continuously stratified systems, often flows can be modeled as a composition of discrete layers.

In this study, a two-layer counter-current (exchange) flow system over a submerged sill is investigated. The main objectives are to investigate the effect of higher Reynolds numbers on the interfacial stability between the two layers, the effect of locally enhanced bottom roughness on interfacial waves and the influence of the bottom slope on hydrodynamic instabilities at the interface. The approach is based on physical experiments using optical measurement techniques and numerical analysis by means of linear stability analysis.

Different initial experimental conditions are investigated, distinguishing the case of hydrostatic disequilibrium and the case of a global pressure-balanced state. The experiments demonstrate that it is impossible to generate experimentally a purely baroclinic exchange flow, without invoking additional barotropic forcing. A simplified barotropic model, adapted for the experimental conditions, is developed. This model can predict the flow rate oscillations and the related period, revealing good agreement with experimental data. The effect of the unsteady superimposed barotropic forcing on the interfacial wave characteristics is further investigated. Large, two-dimensional, surge-like structures are observed during the experiments, whose generation is related to the flow rate oscillations. By analyzing time series of the velocity data, the length scales of these structures reveal to be comparable to the total water depth and to increase with increasing Reynolds number. Moreover, the entrainment coefficients are larger than the values reported in previous studies of stratified exchange and arrested flows.

Further results of experiments with locally enhanced bottom roughness are presented. Two different types of bottom roughness are investigated, namely fine, dense, short bottom roughness elements and tall, sparse roughness elements. Bottom roughness causes an increased energy dissipation and less inclined interface positions in the experiments. The two roughness types (dense or sparse) control differently the generation and collapsing mechanisms of the large-scale two-dimensional structures at the interface between the two layers. The estimated vertical and horizontal length scales of the two-dimensional structures are reduced in the rough cases compared to the smooth case and the related generation period increases linearly with their size.

On the basis of dimensional analysis, three parameters representing the ratio between the different vorticity production sources (baroclinic production, vortex stretching and bottom benerated turbulence), are defined and compared to the different experiments.

These parameters show that the vorticity production due to bottom friction reduces the acceleration of the flow down the slope and acts as an additional restoring force, together with the buoyancy force. Sparse bottom friction gives the lowest entrainment coefficients which are reduced by 1/3 compared to the smooth case and to the case with dense bottom roughness, where the entrainment coefficients are similar.

A first step in understanding the stability of a two-layer stratified exchange flow down a slope was done by performing a linear stability analysis. As the modeled flow for the stability analysis is strongly idealized, it does not take into account the full portrait of the flow processes occurring at the interface. For this reason, results relative to the stability analysis are not compared to experimental data. Comparisons to experimentally measured quantities are done only for validating the choice of certain definitions for the stability analysis, as for example the velocity profiles and other parameters characterizing the shear layer or density layer thickness. Nevertheless, linear instability does correctly describe the onset and early evolution of infinitesimal perturbations and the results can be used as starting point for a non-linear analysis.

Different numerical methods are tested, namely a pseudospectral method employing Chebyshev Polynomials and a finite difference method. Due to the low convergence of the Chebyshev series, a central second-order finite difference method with an irregular grid is employed. The stability is studied from a temporal frame of reference. Results show that for a given wavenumber, the bottom slope has two effects on the stability: first, it increases the unstable regions as the stationary neutral stability curve is not the stability boundary anymore. Second, it introduces in the unstable region the coexistence of two types of instabilities: Kelvin-Helmholtz instabilities characterized with higher temporal amplification rates and which decrease for increasing Richardson numbers, and Holmboe instabilities, which exist at higher Richardson numbers with a finite, non-zero phase speed. The unstable regions increase for larger slope angles. The slope has thus a similar effect on the stability of the flow as the ratio between the shear and density layer thickness.

Kurzfassung

Natürliche Wasserkörper, die über eine Engstelle oder einen Kanal miteinander verbunden sind, können aufgrund unterschiedlichen Salz- oder Sedimentgehalts oder variabler Temperatur eine veränderliche Dichte aufweisen. Dadurch entstehen Strömungsvorgänge, bei denen große Wassermengen samt ihren chemischen, physikalischen, biologischen Wasserqualitätsmerkmalen ausgetauscht werden. Zahlreiche Beispiele finden sich in der Ozeanographie (Straße von Gibraltar, Bosporus, Nord- Ostsee Verbindung) sowie im Bereich der Binnengewässer (Seen und Wasserreservoirs, die über Engstellen verbunden sind, sowie Seitenarme bei Flüssen). An der Zwischenschicht solcher Austauschströmungen treten Strömungsinstabilitäten auf, wie z.B. Kelvin-Helmholtz Wirbel, die zu einem Massenaustausch bzw. Mischungseffekten zwischen der oberen und der unteren Fluidschicht führen, und dadurch deren Wasserqualitäts- und Strömungseigenschaften beeinflussen. Ziel dieser Arbeit ist es, den Zusammenhang zwischen den maßgeblichen Parametern (geometrische Bedingungen, Wassertiefe sowie Dichteunterschiede) und den auftretenden Strömungsinstabilitäten mit den damit verbundenen Misch- und Transportvorgängen im Detail zu beschreiben. Eine Quantifizierung und Prognose dieser Strömungsphänomene soll mittels physikalischer Experimente unter Verwendung moderner synoptischer Messtechniken, sowie durch hydrodynamische Stabilitätsanalysen erreicht werden.

Diese Arbeit besteht aus drei wesentlichen Elementen: 1) die experimentelle Untersuchung einer geschichteten Austauschströmung unter Bedingungen höherer Reynoldszahlen und 2) unter den zusätzlichen Einfluss von erhöhter Bodenrauheit und 3) die lineare Stabilitätsanalyse einer Austauschströmung unter den Einfluss einer räumlichen Beschleunigung. Im Folgenden werden diese drei Elemente zusammengefasst.

1. Einfluss höherer Reynoldszahlen. Unterschiedliche Anfangsbedinugungen für das Experiment einer dichtegeschichteten Austauschströmung wurde zunächst untersucht und diskutiert. Die erste dieser Möglichkeiten besteht darin, das selbe Wasserniveau im rechten und linken Reservoir herzustellen. Unter dieser Konfiguration, besteht ein anfängliches Ungleichgewicht, da die resultierende hydrostatische Kraft auf der Trennwand auf der Salzseite größer ist als die resultierende Kraft auf der Frischwasserseite: das führt zu einer pulsierenden barotropischen Strömung, nachdem die Trennwand entfernt wird. Um diesen Effekt auszugleichen kann die Wassertiefe im frischen Wasserbehälter vor Beginn des Experiments erhöht werden, um damit ein globales Kräftegleichgewicht zu erzielen. Nach dem Ziehen der Trennwand, erzeugt der Niveauunterschied eine barotropische Welle, die der baroklinische Strömung überlagert ist. In beiden Fällen also, ist die Strömung durch eine Überlagerung auf eine baroklinischen Strömung, einer barotropischen Strömung charakterisiert. Experimentelle Ergebnisse zeigen, dass es unmöglich ist, einen ausschließlich baroklinischen Austausch experimentell zu erzeugen. Zur Beschreibung dieses barotropischen Effekts wurde ein vereinfachtes analytisches Modell entwickelt, angepasst

an die experimentellen Bedingungen. Mit Hilfe dieses Ansatzes lassen sich nun der Nettodurchfluß sowie die Periode der barotropischen Schwingungen voraussagen. Schließlich wurde dieser Ansatz mit experimentellen Daten aus dieser und anderen Untersuchungen validiert. Farbvisualisierungen zeigten, dass nicht nur klassische Kelvin-Helmholtz Wirbel, sondern auch interne Wellen, in Form von pulsierenden Salzkeilen an der Zwischenfläche generiert werden. Diese Salzkeile werden erzeugt, um das barotropische Gleichgewicht wieder herzustellen. Diese Art der Instabilität wurde bereits in anderen Untersuchungen beobachtet, konnten aber bislang nicht physikalisch erklärt werden und wurden meist als normale Instabilität ähnlich den Kelvin-Helmholtz-Wirbeln interpretiert. Es konnte gezeigt werden, dass die Erzeugungsfrequenz dieser Instabilitäten nahe der berechneten Frequenz der barotropischen Schwingungen in der Durchflussrate liegen. Weitere fünf Experimente sind zur Analyse der pulsierenden barotropischen Strömung und deren Auswirkung auf die Entwicklung der Scherschicht durchgeführt worden.

Spektren der Geschwindigkeitsfluktuationen besitzen im niedrigen Frequenzbereich eine Steigung von $-7/3$, während im hohen Frequenzbereich die Steigungen bei $-5/3$ liegen. Diese unterschiedlichen Steigungen deuten auf das Zerfallen zweidimensionaler großskaliger Strukturen in Strukturen mit höherer Frequenz hin. Die unterschiedlichen spektralen Verteilungen der Geschwindigkeitsfluktuationen in der horizontalen und vertikalen Richtung deuten weiterhin auf anisotrope Turbulenz im niederfrequenten Bereich hin, das stärker ausgeprägt ist für die Experimente mit höhererer Reynoldszahl. Mittlere Geschwindigkeitsfelder zeigen eine zunehmende Scherschichtdicke mit zunehmender Richardsonzahl. Das Verhältnis $u'/\Delta U$, das die Reaktion der Geschwindigkeitsfluktuationen durch den Geschwindigkeitsunterschied auf Grund der Beschleunigung entlang der Schwelle beschreibt, zeigt eine allgemeine Zunahme in Längsrichtung. Diese Ergebnisse stehen im Gegensatz zu Beobachtungen von Koop and Browand [1976] und Pawlak and Armi [2000]. In diesen beiden Untersuchungen wird von einer Abnehmrate in Längsrichtung von $x^{-3/4}$ berichtet.

Die vertikalen und horizontalen Längenskalen der beobachteten großskaligen Strukturen wurden mit der Zeitskala, welche mittels Autokorrelationsfunktion der Geschwindigkeitsfluktuationen bestimmt wurde, und der konvektiven Geschwindigkeit $\Delta U/2$ abgeschätzt. Die horizontalen und vertikalen Längenskalen zeigen eine allgemeine Zunahme mit zunehmender Wassertiefe, gleichzeitig jedoch eine schwache Abnahme mit zunehmender Auftriebsbeschleunigung bei $Re_* = \sqrt{g'H}H/\nu < 60,000$. Die so definierte Reynoldszahl hat sich als Schlüsselparameter für die Beschreibung der Anfachungsrate der großskaligen Strukturen erwiesen. Die folgende Beziehung zwischen dieser Reynoldszahl und der Längenskalen wurde vorgeschlagen: $L_I/H = \kappa Re_*$ wobei κ $0.9 \cdot 10^{-5}$ für die horizontalen Skalen und $0.33 \cdot 10^{-5}$ für die vertikalen Skalen beträgt. Ein ähnliches Verhalten der Längenskalen wurde auch in Abhängigkeit der Richardsonzahl beobachtet. Allerdings ist hierbei die Abweichung zwischenVorhersage und experimentellen Daten höher. Einmischungsraten zeigen höhere Werte als in den meisten Studien über Zweischichtströmungen (Pawlak and Armi [2000], Morin and Loewen [2004]), bei gleicher Richardsonzahl. Für größere Reynoldszahlen, steigen auch die Einmischungsraten, was überraschend ist, denn normalerweise passiert das Gegenteil. Das kann damit erklärt werden, dass die Erhöhung der Reynoldszahl durch die Erhöhung der Wassertiefe, und somit einer verstärkten barotropischen Strömung gekoppelt ist. Eine stärkere barotropische Strömung erzeugt größere Salzkeile an der Zwischenfläche. Diese großskaligen Wirbelstrukturen dominieren im vorliegenden Strömungszustand das Einmischungsverhalten. Das Verhältnis zwischen der Einmischungsgeschwindigkeit und den mittleren Geschwindigkeitsfluktuationen, aufgetragen gegen die in Pawlak and Armi [2000] definierte Richardsonzahl, zeigt gute Übereinstimmung mit dem Modell von Cotel and Breidenthal [1997], mit einer Abhängigkeit von Ri^{-1}.

2. Einfluss erhöhter Bodenrauheit. Experimentelle Ergebnisse von Versuchen mit lokal erhöhter Bodenreibung werden im Weiteren präsentiert. Die Daten dieser Experimente wurden mit kombinierten PIV/PLIF Messungen erfasst. Zwei Rauheitstypen sind in den Experimenten eingesetzt worden, nämlich dicht beieinanderliegenden, niedrige Rauheitselemente und zerstreute und höhere Rauheitselemente. Diese Ergebnisse wurden mit deren eines Falles mit glatter Sohle verglichen. In einem Fall wurde künstlicher Rasen als lokale erhöhte Bodenrauheit über der Schwelle verwendet, mit einer mittleren Höhe von $1,6\ cm$ und einer sehr hohen Dichte der Rauheitselemente. Im anderen Fall wurden Kunststoffdübel eingesetzt, mit einer Höhe von $3,5\ cm$ jedoch mit einer verringerten Dichte der Rauheitselemente.

   In solchen Strömungen sind drei unterschiedliche Parameter für die Produktion der Wirbel verantwortlich: eine erste Quelle der Wirbelbildung wird durch die Veränderung der Wassertiefe über die Schwelle gegeben. Eine zweite Ursache sind baroklinischerzeugte Wirbel, die durch die Neigung der Scherschicht gegeben ist. Drittens, trägt die erhöhte Bodenrauheit zur Turbulenz bei, die als zusätzliche Quelle der Turbulenzproduktion angesehen werden kann. Durch Dimensionsanalyse der Wirbelgleichung, wurden drei Parameter definiert, um den Beitrag jeder Turbulenzquelle zu analysieren. Das Verhalten der drei Wirbelparameter hat gezeigt, dass die Bodenrauheit zwei wichtige Ursachen für die Stabilisierung der Scherschicht darstellt: erstens verringert sie die Beschleunigung über die Schwelle und damit die Geschwindigkeit in der unteren Schicht. Zweitens erhöht sie die turbulente Energieübertragung zwischen der unteren Schicht und der Scherschicht, mit dem Ergebnis, dass in einem Experiment die großskaligen Wirbelstrukturen sich nicht mehr entwickeln oder, im anderen Fall, dass sich kein Nachlauf hinter der Strukturen bildet.

   Der Vergleich mit der Theorie zur internen Hydraulik hat gezeigt, dass die Rauheit die Position der Scherschicht nach oben verschiebt und die Neigung in Längstrichtung reduziert. Die interne Energie der im Fall der erhöhten Bodenreibung weißt hohe Verluste in der Beschleunigungszone der Schwelle auf und kleinere Werte der internen Froudezahl im Vergleich zu den theoretischen Vorhersagen. Dies hängt einerseits mit den erhöhten Scherschichtlagen zusammen aber auch mit den geringeren Geschwindigkeiten in beiden Schichten im Fall der erhöhten Bodenreibung.

   Die Struktur der Strömung wurde analysiert, wobei die physikalische Beschreibung der Salzkeilen an der Scherschicht über der Schwelle und eine Abschätzung der Längenskalen der beobachteten Strukturen durchgeführt wurde. Auch durch die Spektralanalyse der Geschwindigkeitsfluktuationen, ist die Verkleinerung der Wirbelgröße nachgewiesen worden. Die unterschiedlichen Bodenrauheitselemente zeigen bezüglich der Scherschichtwellen ein unterschiedliches Verhalten: im Falle einer dichten Bodenrauheit, sind die Geschwindigkeitsgradienten im Vergleich mit der weniger dichten Bodenrauheit wesentlich ausgeprägter. Im ersten Fall ist die Geschwindigkeit in der rauhen Zone fast null, steigt dann rasch in vertikaler Richtung auf 5% der Bezugsgeschwindigkeit an. Im Falle einer geringeren Dichte der Rauheitelemente, passiert die gleiche Geschwindigkeitszunahme in einer höheren Schicht sodass geringere Geschwindigkeitsgradienten auftreten. Dies weist darauf hin, dass die Interaktion zwischen der Grenzschicht und der Scherschicht im Falle von dichtenRauheitselementen viel intensiver ist und die zweidimensionale Turbulenz, die an der Scherschicht wegen der superponierten barotropischen Strömung erzeugt wird, wesentlich schneller in dreidimensionale Turbulenz zerfällt.

Das hat auch unterschiedliche Auswirkungen auf die Einmischungsraten: niedrigere Einmischungsraten sind nur für das Experiment mit der niedrigen Dichte der Rauheitselemente beobachtet worden. Im Fall einer glatten Sohle, ist die Hauptquelle der Einmischungsrate der Nachlauf hinter den grosskaligen, barotropisch generierten Wirbelstrukturen. Wenn die Rauheitselemente sehr dicht sind, kommt der Hauptbeitrag für die Einmischungsraten von der Bodenrauheit, die eine sehr intensive Interaktion zwischen der unteren Grenzschicht und der Scherschicht erzeugt. Wenn die Rauheitselemente eine niedrigere Dichte haben, ist der Hauptbeitrag für die Einmischungsrate wieder die Bodenrauheit, jedoch ist in diesem Fall einerseits die Interaktion zwischen der Grenzschicht und der Scherschicht niedriger und andererseits wird die Nachlaufbildung der Salzkeile verhindert, so dass auch in diesem Fall niedrigere Einmischungsraten beobachtet wurden.

3. Lineare Stabilitätsanalyse. Mit Hilfe der numerischen linearen Stabilitätsanalyse konnte eine allgemeine Beschreibung der Stabilität einer dichtegeschichteten Austauschströmung erzielt werden. Vorrangig wurden dabei der Einfluss der räumlichen Beschleunigung untersucht. Die Austauschströmung wurde mit einem hyperbolischen Tangensprofil für die Geschwindigkeitprofile und Dichteprofile approximiert.

   Unterschiedliche numerische Methoden wurden getestet, nämlich eine Spektralmethode, die auf Chebyshev-Polynomen beruht und eine Finite-Differenzen-Methode. Aufgrund hoher Rechenzeiten und relativ geringer Qualität der Ergebnisse, wurde eine zentrierte, Finite-Differenzen-Methode zweiter Ordnung angewendet , in der ein unregelmäßiges Gitter zum Einsatz kam. Die Stabilität wurde zeitlich untersucht. Aus diesen Ergebnissen hat sich ergeben, dass für eine gegebene Wellenzahl, die Bodenneigung zwei Effekte auf die Stabilität der Strömung erzielt: erstens erhöht sie die instabilen Regionen, da nun die stationäre Nullstabilitätskurve nicht mehr die Stabilitätsgrenze ist. Zweitens koexistieren in den instabilen Regionen zwei Arten von Instabilitäten: Kelvin-Helmholtz-Instabilitäten, gekennzeichnet von einer höheren zeitlichen Wachstumsrate, die sich aber bei erhöhter Richardsonzahl verringert. Weiterhin wurden auch Holmboe Instabilitäten nachgewiesen, die bei höheren Richardsonzahlen vorhanden sind und eine abnehmende Phasengeschwindigkeit aufweisen.

Zusammenfassend wurde durch diese Arbeit nachgewiesen, dass es unmöglich ist experimentell eine reine baroklinische Strömung zu erzeugen, ohne zusätzliche barotropische (instationäre) Effekte hervorzurufen. Diese pulsierenden barotropischen Wellen erzeugen großskalige Wirbelstrukturen an der Zwischenschicht, die die Einmischungsraten und die Stabilität der Zwischenscherschicht signifikant beeinträchtigen. Die sich daraus ergebenden Strömungen sind in der Natur sowohl in Binnengewässern als interne Seichen zu beobachten, als auch in der Ozeanographie und der Metereologie, wo metereologische und/oder astronomische Gezeiten eine Rolle spielen. Weiterhin wurde gezeigt, dass die Bodenrauheit, je nach Typ, die Einmischungsraten und die Stabilität der Scherschicht unterschiedlich beeinflussen kann. Die lineare Stabilitätsanalyse hat schließlich gezeigt, dass die Bodenneigung eine destabilisierenden Effekt aufweist, indem eine zweite Art von Instabilität generiert wird, die durch eine Phasengeschwindigkeit characterisiert ist. Ergebnisse aus der Stabilitätsanalyse wurden jedoch nicht mit den experimentellen Daten verglichen, da die untersuchte experimentelle Strömung von Instabilitäten dominiert ist, welche sich mit einer linearen Stabilitätsanalyse nicht vorhersagen lassen: die Instabilitäten, welche während der experimente beobachtet wurde, sind das Ergebnis einer Wechselwirkung zwischen barotropische und baroklinische Effekte. Diese können durch die stark vereinfachenden Annahmen für die lineare Stabilitätsanalyse in den ergebnissen nicht wiedererkannt werden.

Acknowledgements

This dissertation was supported by the German Science Foundation (DFG-Ji18/02).

My deepest gratitude is due firstly to my advisor, Professor Jirka, whose guidance has always included just the right amount of direction while allowing me just the right amount of autonomy.

I owe thanks to all my committee members, especially to Prof. David Zhu, always accessible for stimulating discussions. I am also deeply grateful to Scott Socolofsky for his patience, guidance and support for the numerical part of this thesis.

I am deeply indebted to Andreas Rummel and Gregor Kühn, who were always available for any problems in the laboratory during both the planning and running phase of my experiments and for any discussions sheared with me. I thank Volker Weitbrecht for the constructive fights we had, Martin Detert for the flowers and Herlina for the discussions about science and about women life in research.

I am deeply indebted also to Nikolai Stache, whose independent workwise saved me a lot of time and patience.

I owe thanks to Manuel Garcia Villalba, not only for the precious discussions on science, but also for listen to me in difficult situations. I also thank John Fenton, who was always available for any discussion on mathematical issues.

I thank the department leaders (Prof. Jirka, Prof. Lang and Dr. Weitbrecht) for the given time and the comprehension in the summer 2006.

Special thanks go to Tilman Friz for always supporting me in my ambitions and for all the things that can not be put in words.

Finally, I owe thanks to my family and to Irene Bellucco for their love and for always supporting my unsatiable thirst of seeing the world outside Italy.

Karlsruhe, 11.07.2007

Publications

The content of this thesis is based on the following publications which have been published or have been submitted for publication during the course of the PhD program:

- Negretti ME, Zhu DZ and Jirka GH (2007) *Barotropically induced interfacial waves in stratified two-layer flows over a sill*, J.Fluid Mech. (in press).

- Negretti ME, Zhu DZ and Jirka GH (2007) *The effect of bottom roughness in two-layer flows down a slope*, submitted to J. Phys. Ocean.

- Negretti ME, Socolofsky SA and Jirka GH (2007) *Linear stability analysis of spatially accelerating stratified shear flows*,
  Proceedings of 32nd Congress of the International Association of Hydraulic Enginnering (IAHR), Venice, July 1–4, 2007.

- Negretti ME and Jirka GH (2006) *Hydrodynamic Instabilities and mixing and entrainment in two-layer stratified exchange flows over a submerged sill*
  Technical Report (DFG Report Nr. 831), Project Nr. Ji18/12-1, 16 pages.

The dissertation author was the primary investigator and first author of these manuscripts.

Cinque volte racceso e tante casso
lo lume era di sotto da la luna
poi che 'ntrati eravamo nell'alto passo,
quando n'apparve una montagna, bruna
per la distanza, e parvemi alta tanto
quanto veduta non avëa alcuna.
Noi ci allegrammo, e tosto tornò in pianto;
che de la nova terra un turbo nacque
e percosse del legno il primo canto.
Tre volte il fé girar con tutte l'acque;
a la quarta levar la poppa in suso
e la prora ire in giù, com'altrui piacque,
infin che 'l mar fu sovra noi rinchiuso.

(From *The last trip of Odysseus trough the Strait of Gibraltar* by Dante Alighieri [Cerchio VIII, bolgia 8^a, IF XXVI vs. 130–142.])

Contents

List of Figures

List of Tables

List of Symbols

Latin

A	area	m^2
a	attenuation coefficient	m^{-1}
B	width of the connecting channel	m
C	concentration	$\mu g/l$
E	internal energy	m
F	Froude number	$-$
F_s	hydrostatic force of salt water	N
F_w	hydrostatic force of fresh water	N
f	frequency (or Darcy-Weisbach friction parameter)	Hz
G	internal (composite) Froude number	$-$
g	acceleration due to gravity	m^2/s
g'	buoyant acceleration	m^2/s
H	total water depth	m
h	local water depth	m
J	bulk Richardson number	$-$
k_s	friction coefficient	m
L	right length of the sill from sill crest	m
L_C	length of connecting channel	m
L_I	integral length scale	m
N	natural buoyancy frequency	Hz
N_g	number of grid points	$-$
p	pressure term	Pa
Pr	Prandtl number	$-$
q	flow rate per unit width	m^2/s
T	oscillation period	s
T_A	acceleration parameter	$-$
t	time	s
TB	mapped Chebyshev Polynomials	s
U	mean velocity	m/s
u, v, w	local velocity components in x, y, z-direction	m/s
u_e	entrainment velocity	m/s
u_*	shear velocity	m/s
u'	longitudinal component of the Root-Mean-Square-Velocity	m/s
R	ratio between velocity and density shear layer thickness	$-$
Re	Reynolds number	$-$
Ri	Richardson number	$-$

S	sill slope	$-$
Z	vorticity ratio parameter	$-$

Greek

α	wave number and spatial amplification rate of disturbance	$-$
α_e	entrainment rate	$-$
$\alpha_\star$	roughness coefficient	m
β	frequency and temporal amplification rate of disturbance	$-$
γ	coefficient of expansion of the stratifying agent	$-$
Δ	difference	$-$
δ	velocity shear layer thickness	m
δ_ν	vorticity thickness	m
δ_ρ	density layer thickness	m
ϵ	offset of density interface from velocity interface	$-$
η	density layer thickness	$-$
κ	diffusivity of the stratifying agent	$-$
λ	Chebyshev mapping parameter	$-$
μ	molecular viscosity	$kg/(ms)$
ν	kinematic viscosity	m^2/s
θ	sill slope	$radians$
σ	normalized density profile	$-$
ϕ	amplification rate of disturbance	$-$
ψ	stream function	s^{-1}

Subscripts

1	certain quantity in the upper layer
2	certain quantity in the lower layer
x, y, z	spatial derivative

1 Introduction

Stratified flows and the related flow processes attract considerable interest and are a widely studied aspect of fluid mechanics. A special class of stratified flows is represented by the equal exchange flow, in which the two layers are flowing with the same flow rate in opposite directions (see figure 1.1). Such flows are ubiquitous for example in environmental and geophysical flows, like interconnections between two water bodies of different density (Farmer and Armi [1999]) and in the atmosphere (Assaf and Hecht [1974]) between two valleys connected by a pass (see figure 1.2) and in architectural flows (Linden [1999]), between the in- and outside of buildings through openings (windows, vents), driven by temperature differences.

In environmental and geophysical flows stratified exchange flows typically occur at bottlenecks in the ocean where large oceanic basins are connected by a small channel through which pronounced water masses containing various water quality properties (e.g. chemical, physical and biological constituents and species) can be exchanged. Examples in the oceanographic context include the Strait of Bosphorus connecting the Mediterranean sea with the Black sea, the Faroe Bank channel connecting the Greenland-Iceland-Norwegian Seas to the North Atlantic (Sherwin and Turrell [2005]) and the Strait of Gibraltar, connecting the Mediterranean sea with the Atlantic ocean (Sannino and Artale [2002]).

Since the beginning of the 20th century the Strait of Gibraltar has been a favorite place for oceanographic studies. The dominant oceanographic feature of the Strait is a two-layer inverse estuarine circulation that is driven by an excess of evaporation over precipitation and river discharge into the Mediterranean (see figure 1.3 (a)). Recent field experiments in the Strait have kindled a renewed interest in two-layer hydraulics and its role in controlling the mass transport between the Mediterranean and the Atlantic sea. The Mediterranean sea is a basin in which a wide range of oceanic processes and interactions of global interest occur. Higher evaporation over the Mediterranean Sea produces dense, salty water that outflows through the Strait of Gibraltar and can have an impact on the general circulation of the global ocean and particularly on the formation of deep water in the North Atlantic. One effect of the outflow from the Mediterranean

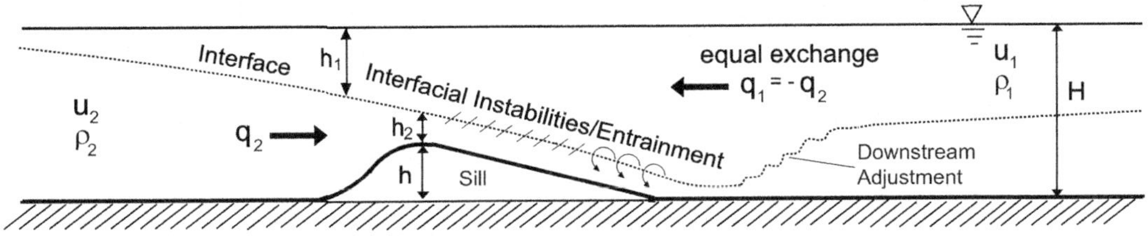

Figure 1.1: Definition sketch of a two-layer equal exchange flow over a sill.

Figure 1.2: An example of stratified exchange flow in the atmosphere: fog sea in a Rinevalley, flowing into another valley due to temperature differences.

is to make the Atlantic Ocean saltier than either the Pacific or Indian Oceans.

The Gibraltar Strait controls the entire fresh water budget of the Mediterranean sea and for this reason there has been both experimental and theoretical interest in the physics of the Strait and on monitoring the heat and mass flux down the bottom sill in the Strait (see figure 1.3 (b)). The existence of this strait makes the Mediterranean sea one of the few regions in the world's oceans where the advective heat and salt flow is known with high accuracy.

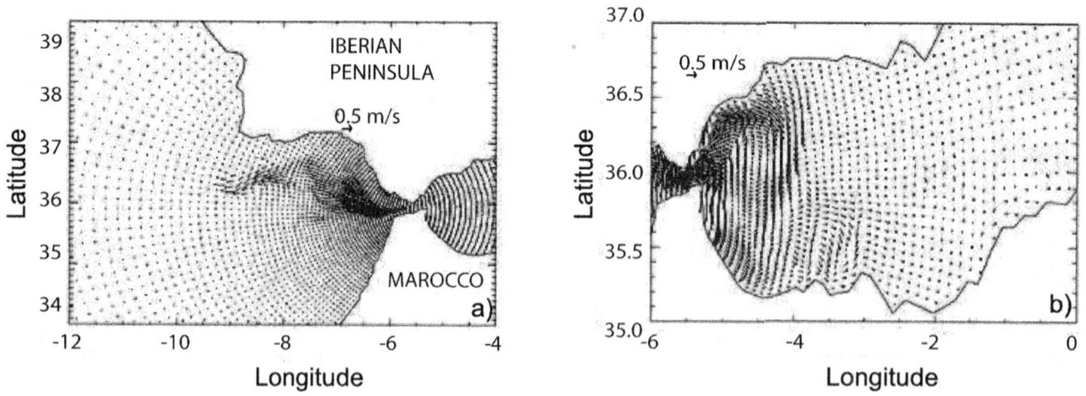

Figure 1.3: Velocity distribution in the Strait of Gibraltar. a) Velocity field of the Mediterranean outflow and b) of the Atlantic inflow. From Sannino and Artale [2002].

Geophysical exchange flows can also be driven by a small density difference between a lake and its embayment as a result of temperature differences, such as the exchange flow between different portions of a lake as in Lake Lucerne [Wuest and Schurter, 1988, Aeschbach, 1996] and the exchange flow between the Lake Ontario and Hamilton Harbor through the Burlington Ship Canal [Hamblin and Lawrence, 1990]. Predicting the rate of this flow exchange and the mixing and fate of the exchanged inflows into their neighboring water bodies is important in environmental and ecological studies such as water quality modeling and pollution remediation.

Earlier studies have shown that the rate of the exchanges are controlled and/or influenced by the generation of hydrodynamic (interfacial) instabilities at the interface of the two layers [Pratt, 1987, Zhu and Lawrence, 2000]. Significant flow entrainment is also reported when the heavier fluid in the lower layer leaves the connecting channel and flows down a slope into the neighboring water body [Morin and Loewen, 2004], similar to the turbidity currents studied by Parker and Fukushima [1987]. Morin and Loewen [2004] reported that the generation and break-up of large scale Kevin-Helmholtz instabilities at the interface of the two layers are mainly responsible for the entrainment of the ambient lighter fluid into the heavier lower layer. This flow entrainment thus modifies the density of the inflow and its eventual mixing and fate. Various review articles on mixing in stratified lakes and oceans [Thorpe, 1985, Fernando, 1991, Wuest and Farmer, 2001] document that hydrodynamic instabilities are largely responsible for the vertical mixing in the interior of oceans and lakes. A fundamental study on hydrodynamic instabilities is important for modeling the vertical transport of heat, oxygen, nutrients and pollutants in lakes and oceans, as density stratification limits the vertical movement.

Oceans represent the biggest reservoir of CO_2. The vertical transport rate of CO_2 in the stratified ocean water directly influences the absorption rate of CO_2, which needs to be quantified in order to properly model the balance of CO_2 in the atmosphere, an important issue in modeling global warming. In temperature stratified lakes and reservoirs, the stratification reduces the vertical movement of the water and prevents the aeration of cool water in the hypolimnion [Patterson and Imberger, 1984].

Thus it could result in fish kills when the dissolved oxygen in the hypolimnion is depleted.

The entrainment induced by the heavier inflow along the bottom slope is also important in many engineering problems, for example, river water of high sediment concentration flowing into a reservoir [Alavian, 2002], seawater intrusion into a lake [Dallimore and Ishikawa, 2001] or a fjord [Farmer and Armi, 1999]. Flowing along the bottom, the dense water entrains the lighter, oxygen rich ambient water and carries it into deeper layers. The fate of the dense water depends on the amount of the entrainment (i.e. dilution) as well as the type of stratification in the receiving water body. Previous research has concentrated on quantifying the mixing by the use of bulk entrainment coefficients rather than looking at the mechanisms of the mixing. As a result, predictions of the rate of entrainment can vary by more than 10 times [Alavian, 2002].

The quantitative understanding of shear-induced instability and subsequent mixing is important in predicting the vertical transport of heat, oxygen, nutrients and pollutants in inland water bodies, oceans, and the atmosphere (see figure 1.2). This has relevance to many branches of science and engineering, such as hydrodynamics, meteorology, oceanography, and engineering. Farmer and Armi [1999] hypothesized that these instabilities and

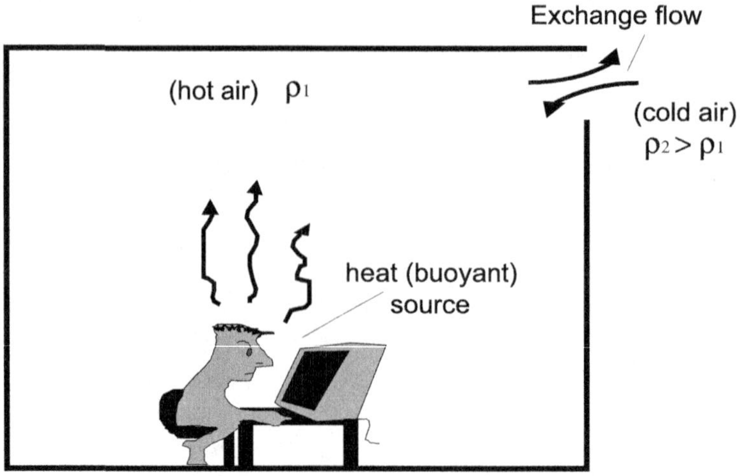

Figure 1.4: Sketch of a two-layer exchange flow between the in- and outside of a building with internal heat source induced by a temperature difference (see Linden [1999]).

mixing results in the establishment of the down-slope hydraulic flow in the Knight Inlet, British Columbia, Canada. In atmospheric sciences, it has been estimated that instability and mixing causes a 50% increase in the drag of atmospheric flows compared to the case in which mixing is neglected and thus is substantial for atmospheric flow modeling and weather forecasting [Afanasyev and Peltier, 2001]. Moreover, the study of atmospheric exchange flows is important for predictions of the air quality and circulation in cities located in valleys, as shown in figure 1.2, where a fog sea in a Rinevalley is flowing into another valley due to temperature differences.

Another example of exchange flows in the geological/geophysical context is given by the exchange between the relatively dense fluid in a volcanic crater and the light fluid contained in a deep magma reservoir (magma chamber), as for example the fluxes of SO_2 and other volatiles from lava lakes in volcanic craters. The evaluation of the amount of magma required to come to the surface to supply the SO_2 flux can be calculated from the known amount of SO_2 dissolved in and transported by the magma. This value of magma flux has shown to exceed over four orders of magnitude the actual erupted flux [Francis and Stevenson, 1993, Wallace, 2003]. The explanation for this discrepancy was given by Huppert and Hallworth [2006]: the magma relatively rich in dissolved SO_2 rises from the magma chamber. As the magma rises, the pressure decreases, which leads to degassing of the SO_2. The degassed SO_2 travels mainly at the same velocity as the rising magma, but when the magma reaches the surface the gas can escape. This leaves the magma with an increased density, which is further increased by the cooling that occurs, and also by crystallization. The relatively heavy magma then sinks back down the connecting conduit (generating a bi-directional exchange flow), which can continue transporting SO_2 to the surface without eruption of any magma. The rate of flow in the conduit has shown to be partly determined by the change in density which is itself set by the heat transfer between the lava lake and the atmosphere. Exchange flows are also common in architectural fluid mechanics, a research field which is attracting more and more interest from the fluid-dynamic community. To improve the internal quality of buildings, particularly in terms of comfort and temperature, there has been an increasing move toward the use of air conditioning in modern buildings. This has undesirable energy implications and leads to high carbon dioxide emissions. As a result there has been a reawakened interest in the use of natural ventilation to provide a better indoor quality, both in commercial and industrial buildings, where increasingly strict environmental and health regulations concerning air quality have to be met. Results of research carried out in the field of natural ventilation in buildings show that these techniques reduce by 20% the energy consumption for air-conditioning systems. Natural ventilation uses the freely available resources of the wind and thermal energy that is a result of solar and incidental heating of the building. Although these resources are free, they are difficult to control. An important issue is to provide the necessary control mechanisms to develop the required indoor air quality. Ventilation is essentially the flow of air between the inside and the outside of a building. This flow occurs through vents, traditionally windows, but more and more through purposely designed, controlled openings not necessarily used for introducing light. Herein, the air flows generated are the result of wind-induced forces on one hand, and of buoyancy (due to temperature differences, also called stack-driven flow), on the other hand, and/or even a combination of them. Much of the early work on the interaction of wind with buildings has been concerned with aerodynamic loading [Owen, 1971]. To include effects of viscosity, the group at Cambridge University have developed the methodology of small-scale modeling using water as the working fluid [Linden and Smeed, 1990, Baker and Linden, 1991]. Buoyancy forces are produced by salinity differences within the fluid. Single sided ventilation, which occurs when there is a single opening into a space and in which exchange flows develop, was first studied experimentally by Brown and Solvason [1962a,b]. This situation is schematically depicted in figure 1.4. The hydraulic of the exchange flow trough the opening in single-sided ventilation was investigated by Dalziel and Lane-Serff [1991]. Wilson and Keil [1990] carried out experiments on the exchange flow through a window in a heated, sealed room of a test house and found disagreements for the discharge coefficient found in the previous studies. This was dictated by the superimposed wind effect. The exchange flow trough the opening, if it is not in a vertical, but in a inclined wall, was investigated by Brown and Solvason [1963], Epstein [1988], Keil [1991] and Davies [1993].

Thus, the study of stratified exchange flows and the understanding of the flow phenomena related to such flows is important in many branches of science and finds a lot of applications in industrial and environmental problems.

1.1 Purpose

Several numerical experiments have recently been reported in studying two-layer exchange flows: the exchange flow through the Strait of Gibraltar [Sannino and Artale, 2002] and through the Danish Strait [Kase and Oschlies, 2000]. While in these studies it was possible to model the exchange flow, the accuracy of these numerical results is hard to judge, due to the lack of good field data. It has been noticed, though, that these models assume a hydrostatic pressure distribution and ignore interfacial entrainment. As a result, the obtained interface appears to be stable to interfacial instabilities, which is in contradiction to laboratory and field observations. A prudent approach might be to calibrate and validate the numerical models using laboratory measurements before these models can be applied to real flow situations. The majority of previous works studied stratified two-layered flows for relative low Reynolds numbers, and thus the results can not be applicable directly to natural flows as the flow behavior and the related phenomena are expected to change if the flow is characterized by higher Reynolds numbers. The first main purpose of this study is to investigate the effect of higher Reynolds numbers as compared to previous studies on the interfacial waves and on mixing and entrainment processes.

In natural water bodies, in the oceans and in the atmosphere, bottom roughness is ubiquitous and is a key parameter in influencing the flow behavior and the related entrainment and mixing processes. Indeed, most of the previous studies on stratified exchange flows dealing with boundary layer effects focused their attention on the subsequent effects on the hydraulics of these flows rather than analyzing their effect on the interfacial stability. The second main purpose of this work is to study boundary layer effects caused by local enhanced bottom friction on the interfacial wave activity and on mixing and entrainment processes.

Under these conditions of higher turbulence and bottom roughness, this study will deliver more representative results for the natural flows as they occur in the oceans, lakes and reservoirs as well as in the atmosphere.

Earlier studies on hydrodynamic instabilities have shown that the quantity of exchanged water masses and the related entrainment are controlled and/or influenced by the generation of hydrodynamic instabilities at the interface of the two layers. It is still not clear which role plays the bottom slope on their generation and developing mechanism. A fundamental and separate understanding of the effect of a bottom slope is needed in order to correctly predict the onset and type of hydrodynamic instability at the interface, as topographical obstructions are common in environmental and geophysical flows.

1.2 Approach

To study the stratified exchange flow down a submerged sill, physical and numerical experiments will be carried out in order to determine velocity and density fields and to investigate the stability of the flow. A fundamental understanding of the interfacial processes that lead to the entrainment is gained through detailed laboratory experiments, which are used for developing simplified models and parameters for prediction and management of the quality of our inland and coastal waters. Laboratory experiments are conducted using Particle Image Velocimetry (PIV) to obtain velocity fields and fluctuations and Planar Laser induced Fluorescence (PLIF) to obtain concentration fields. The physical experiments will also provide the necessary data to validate eventual numerical experiments.

A fundamental understanding of the effect of an inclined bottom (downslope acceleration) on the generation and developing mechanism of hydrodynamic instabilities, such as Kelvin-Helmholtz and Holmboe instabilities (see figure 2.2), at the interface between the two-layers will be achieved by means of linear stability analysis. Although linear stability analysis is only valid for a very short time before nonlinear effects become important, it does correctly describe the onset and early evolution of infinitesimal perturbations and the results can be used as a guide, indicating which parameters may produce instabilities of interest. Moreover, the results from linear stability analysis can be used as starting perturbations for the analysis of the nonlinear problem. Experimental data will be used for comparison in order to justify certain definitions and parameter choices for the linear stability analysis.

In this study, we will focus on the conditions for the generation of hydrodynamic instabilities at the interface of two-layer flows, the growth and eventual break-up of these instabilities, as well as the induced flow entrainment and mixing. While the study focuses on the hydrodynamic instabilities in two-layer flows, the results are generally applicable to any sheared stratified flow.

1.3 Outline

The thesis is organized as follows: in chapter 2, a general literature review of stratified, two-layer flows is given. Chapter 3 explains in detail the employed experimental facility and techniques. In chapter 4 first the experimental starting conditions are discussed and the main properties of the baroclinic exchange flow with superimposed barotropic oscillation are described. The experimental results relative to higher Reynolds numbers are presented in chapter 5. Chapter 6 presents results relative to the effect of local enhanced bottom roughness. The governing equations, the numerical solution method and the results relative to the linear stability analysis are presented in Chapter 7. Finally, chapter 8 includes concluding remarks.

2 Literature Review

The flow between two basins of different density (ρ_1 and ρ_2) can take on different forms depending on the relative interplay between barotropic (an overall pressure gradient causing a net flow) and baroclinic (due to the density difference) forcing [Turner, 1968, Baines, 1975]. Two extreme cases are the equal two-layer exchange flow with eventual barotropic effects and the arrested flow with a net motion in the lower (arrested surface wedge) or in the upper (arrested bottom wedge) layer only. These two extreme cases are schematically depicted in figure 2.1. It is important to distinguish between hydraulically controlled arrested wedge flows and gravity currents and intrusions. Arrested (bottom or surface) wedges are simply the high barotropic flow limit of the equal two-layer exchange flow, where they are controlled by the geometry of the environment in combination with hydraulics. However, these flows have typically been considered from the perspective of gravity currents, as discussed by Benjamin [1968], but gravity currents are propagating density fronts, in which there is a local balance between buoyant force and pressure drag. Various authors [Arita and Jirka, 1987, Stefan and Johnson, 1989] have also attempt to analyze arrested wedges by considering a density front arrested by friction. Arita and Jirka [1987] discussed an interfacial frictional model in the context of these arrested density currents and Jirka and Arita [1987] addressed methods for controlling intrusion shape using boundary layer control. In the following, the review is divided in four sections:

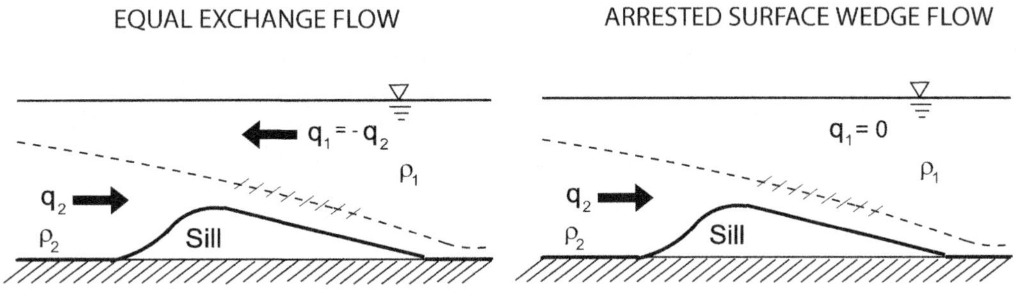

Figure 2.1: Two extreme cases of two-layer stratified flows: the equal exchange (counter) flow, having the same flow rates in the upper and lower layer , but flowing in opposite directions (left) and the highly barotropic limit of the equal exchange flow, the arrested surface wedge flow (right) in which only one layer is moving (here the lower layer). In this study, an equal exchange flow (left) is investigated.

the first one (section 2.1) summarizes past work about the hydraulic theory, the second section (2.2) treats the studies on hydrodynamic instabilities, while the third section (2.3) resumes entrainment studies. A brief summary on the effects of a superposed barotropic flow on the pure baroclinic stratified shear flows is given in section 2.4.

2.1 Two-layer hydraulic theory

Hydraulic theory of two-layer flows was first studied by Schijf and Schoenfeld [1953] who extended classical one-layer hydraulics to two-layer (seawater and freshwater) estuaries. Some unique behaviors of two-layer flows were demonstrated by Long [1954] in his classical experiments of two-layer flow past a bottom sill. Several studies Turner [1968] also examined arrested wedge flows and a density current flowing into a lake.

Armi [1986] studied systematically the hydraulics of two-layer flows. Armi and Farmer [1986] and Farmer and Armi [1986] applied the two-layer hydraulics to bi-directional two-layer flows (i.e. exchange flows) and studied flow regimes as a function of flow conditions, channel geometry and topography, e.g. a channel constriction, a bottom sill, or combinations of a sill and a contraction. In their studies, Armi and Farmer treated the flow as homogeneous layers of inviscid, non-diffusive fluid with a hydrostatic pressure distribution. While these assumptions may not be valid for some flows, the two-layer hydraulic, or internal hydraulic theory, provides the starting point in studying two-layer flows. In two-layer flows, the flow regime is described using the composite (or internal) Froude number G [Armi, 1986] $G^2 = F_1^2 + F_2^2$, where $F_i^2 = u_i^2/g'h_i$ is the densimetric Froude number for layer i, with u_i as the mean horizontal velocity, and h_i as depth of the layer i, see figure 2.1. The reduced gravitational acceleration g' is defined as $g' = g(\rho_2 - \rho_1)/\rho_2$ with ρ_1 and ρ_2 being the density of the upper and lower layer, respectively.

When $G < 1$ (or $G > 1$), the flow is internally subcritical (or supercritical), and when $G = 1$ the flow is internally critical. The location where $G = 1$ is the (internal) hydraulic control. Two-layer flows can have up to two internal controls. Exchange flows with two controls are called maximal exchange flows. Exchange flows with one control are called submaximal exchange flows Farmer and Armi [1986]. Zhu [2002] developed a control curve approach for studying two-layer flows and applied it to identify the locations of controls and flow regimes, as well as to predict exchange flows through a complex channel geometry and topography.

Pratt [1987] reported that due to frictional effects, the location of a control, located originally at the crest of a bottom sill, could be shifted away from the crest point, and for a long channel or strait, the effect of friction cannot be neglected. Lawrence [1993] analyzed theoretically and experimentally a steady two-layer flow past a fixed obstacle and found that an important regime of the flow (namely, approach control flow) cannot be predicted, even qualitatively, using the internal hydraulic theory. Zhu and Lawrence [1998] extended the two-layer hydraulic theory by incorporating the non-hydrostatic pressure distribution due to streamline curvature. The extended theory is able to predict all regimes of the flows studied by Lawrence [1993]. Furthermore, Zhu and Lawrence [2000] extended the theory to include both, friction and non-hydrostatic pressure distribution, and obtained excellent agreement with their experimental measurements of exchange flows through a channel of constant width with a bottom sill. Thus, interfacial friction cannot be neglected in a rigorous dynamical treatment of these flows.

Anati and Thompson [1977] examined the relative importance of frictional and inertial forces in exchange flows in constant-width channels. They classified the dynamic length of the channel by the parameter $\alpha = f_w L_c/H$, where f_w is the bottom drag coefficient, L_c is the channel length and H is the channel depth. If $\alpha \gg 1$ inertial forces can be neglected while if $\alpha \ll 1$, frictional forces can be neglected. The transition case is given by $\alpha \sim 1$ where both contributes are important. Zaremba and Pieters [2003] developed a numerical model to study effects of friction on the steady stratified exchange flow which revealed three viscous flow regimes for a converging-diverging connecting channel and further three flow regimes when a sill is introduced (see also the numerical studies of Assaf and Hecht [1974] and Oguz and Latif [1990]). More recently, Gu and Lawrence [2005] developed a one-dimensional analytical solution to the fully nonlinear problem of two-layer frictional exchange flow within the context of internal hydraulics in a rectangular channel of constant width.

Thorpe [1983] performed experiments of two-layer stratified flows down a slope over a rough floor, where the roughness elements were represented by square bars set at regular intervals. He observed the generation of radiating internal gravity waves at the initial stage of the flow and reported turbulent mixing behind the bars due to flow separation but low interaction with the interface between the two layers. He showed that this turbulent layer spread vertically less rapidly than the internal waves and that the rate of spread depends on the separation of the bars: the nearer the bars, the higher was the interaction between the bottom generated and the interface generated turbulence. This interaction was additionally enhanced by using an array of square cubes instead of the bars. Myrhaug and Slaattelid [1998] studied analytically and numerically the bottom shear stresses and velocity profiles in stratified tidal planetary boundary layer flows by using similarity theory and determined for flow conditions in the rough, smooth and transitional smooth to-rough turbulent regime the maximum bottom shear stress, for a given seabed roughness length, free stream current velocity components, frequency of tidal oscillation, Coriolis parameter and stratification parameter.

More recently, Skyllingstad and Wijesekera [2004] used a three-dimensional large-eddy simulation (LES) model to examine how stratified flow interacts with bottom obstacles in the coastal ocean. Their simulations yielded flow responses ranging from transition flows with relatively high internal wave pressure drag to super-critical flow with relatively small internal wave drag. The cases with high wave drag exhibited strong lee-wave systems and regions of turbulent overturning. Application of bottom drag caused the formation of a bottom boundary layer, which greatly reduced the strength of lee-wave systems in the transition cases. Reidenbach and Genin [2006] performed measurements of velocity and rates of turbulence over a fringing coral reef (Red Sea) to determine the effect of bottom roughness on the flow dynamics and on the mixing in boundary layers and find a good agreement between the existing boundary layer flow theory and the data over rough topography of coral reef (see also the study of Fugate and Chant [2005]).

2.2 Hydrodynamic instabilities

The intense velocity gradients between the two layers in a two-layer stratified flow generate hydrodynamic instabilities, such as Kelvin-Helmholtz (see figure 2.2 (a)) and Holmboe (see figure 2.2 (b)) instabilities. Earlier studies mostly focused on Kelvin-Helmholtz (in the following KH) instabilities as they are commonly observed in the atmosphere [Turner, 1968, Drazin and Reid, 1981]. KH instabilities are characterized by the rolling up of the density interface leading to two-dimensional billows. Sheared density interfaces have been studied analytically since the work of Taylor [1931] and Goldstein [1931]. Their analysis led to an equation describing the stability of inviscid, parallel stratified shear flows.

Holmboe [1962] applied temporal linear stability analysis to piecewise linear velocity profiles with a two-layer density structure. With a fixed limit of the Richardson number, he predicted the occurrence KH instabilities and reported the possible existence of another type of instabilities when the thickness of the density layer is much smaller than the shear layer. This type of instability, named after Holmboe (see figure 2.2 (b)), is characterized as two sets of waves with one layer cusping into the upper layer and the other into the lower layer, and propagating in opposite direction.

Hazel [1972] conducted detailed numerical experiments on the conditions for Holmboe instabilities, and found that Holmboe instabilities can be generated at any weak shear. He examined the stability of flows where the density interface thickness has a finite value and where the stability is characterized by the ratio of the shear layer thickness δ to the density interface thickness η, $R = \delta/\eta$. He found critical values for R below which Holmboe instabilities do not exist. By numerically computing stability diagrams for various values of R, Smyth and Peltier [1988] found the critical value of R to be approximately 2.4. Smyth and Peltier [1989, 1991] studied numerically the transition between the KH instability and Holmboe instabilities, and Smyth and Peltier [1988] investigated finite growth of Holmboe waves.

Lawrence and Redekopp [1991] explained the behavior of asymmetric Holmboe waves (i.e. the two sets of waves having different growth rates and propagation speed) by displacing the density layer from the shear center. Theoretical studies by Haigh and Lawrence [1999] on salinity stratified flows showed that the bulk Richardson number J needs to be less than 0.071 for KH instabilities to be generated, where J is defined as $J = g'\delta/(\Delta U)^2$ with g' as the reduced gravitational acceleration between the layers, δ the thickness of the mixing layer and ΔU the velocity difference between the upper and the lower layer. But, when J is larger than 0.047, Holmboe instabilities are expected to be dominant because they have a larger growth rate.

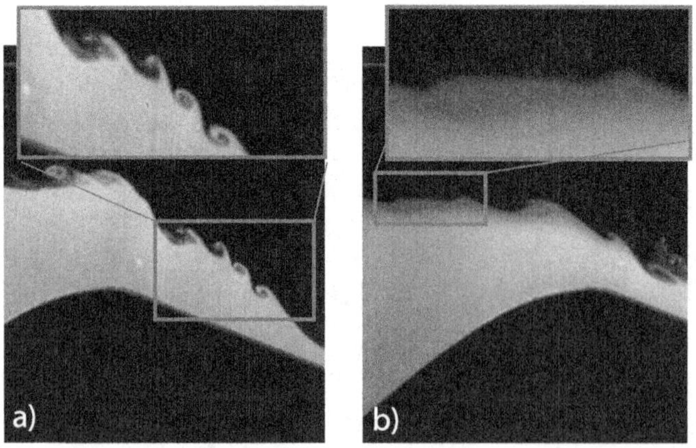

Figure 2.2: Interfacial instabilities generated at the interface in a two-layer exchange flow. a) Kelvin-Helmholtz instability. b) Holmboe instability.

Thorpe [1973] showed KH instabilities in his pioneering tilting tank experiments of two-layer fluids. The importance of KH instabilities in causing vertical mixing in stratified oceans and lakes is well recognized [Fischer, 1979, Thorpe, 1987].

Zhu and Lawrence [2001] reported the occurrence of symmetric Holmboe waves in the subcritical region of their exchange flow experiments. They found that Holmboe instabilities decay when Richardson number exceeds 0.7. Similar results have been reported in the experiments of Strang and Fernando [2001] and Hogg and Ivey [2003].

Morin and Loewen [2004] studied KH instabilities observed in the supercritical region downstream of a sill crest in an exchange flow. Their analysis of interfacial wave activities showed that higher frequency KH waves are continuously being generated in the initial developing flow region. This wave generation stops further downstream. This generation of higher frequency KH waves has not been reported in the literature and is important, because measurements show that the entrainment coefficient of the upper layer fluid into the lower layer is much larger in this region. Their measured entrainment coefficient is found to be several times larger than the prediction of Alavian [2002] for the same Richardson number. This is mainly due to the development and break-up of the large-scale KH instabilities. The entrained fluid modifies the density of the lower layer and eventual spreads into the receiving environment. The measurements of the velocity field by Morin and Loewen [2004] shows, while KH instabilities are generated the shear is substantially weak with J ranging from 0.1 to 0.2.

Pawlak and Armi [2000] also observed KH instabilities in their arrested flows downstream of the sill crest with J ranging from 0.075 to 0.32. These values are larger than 0.071 required for the generation of KH instabilities as predicted by Haigh and Lawrence [1999]. This discrepancy is probably due to significant spatial flow acceleration in the developing region of the flow. Additionally, the proximity of the bottom boundary affects the velocity structure of the lower layer, and might influence the behavior of the KH instability.

The intense velocity gradient at the two-layer interface and the subsequently generated interfacial instabilities cause intense interfacial mixing. Finnigan and Ivey [2000], Finnigan and Ivey [2001] and Finnigan and Lukas [2002] presented a scaling analysis for the effect of diffusivity at the interface and a numerical modeling technique to quantify that effect. As in a turbulent jet, the fast moving layer tends to entrain fluid from the slow moving layer. Gerdes and Farmer [2002] examined the effect of this entrainment mechanism on internal hydraulics for a simple flow configurations where only the lower layer is moving. Morin and Loewen [2004] reported that for supercritical flow downstream of a sill crest, there is significant entrainment induced by the fast moving lower layer. Pawlak and Armi [2000] obtained similar results in their experiments on arrested flows. Further research is needed to incorporate the flow entrainment into the internal hydraulic theory.

Additional literature review on recent stability analysis studies of two-layer stratified flows is given in chapter 7.

2.3 Entrainment

The entrainment and mixing in stratified shear flows have received significant attention due to their importance in environmental problems. Ellison and Turner [1959] studied the problem of buoyant flow of methane gas along the roof of a mine shaft, and examined the entrainment of a developed stratified current along a slope. The entrainment coefficient of ambient fluid into the flowing layer was found to have a power law relation to the bulk Richardson number (defined with the thickness and mean velocity of the gravity current). Subsequent research in this area focused on two types of two-layer stratified flows: those on a horizontal plane and those down a slope. Christodoulou [1986] examined earlier experimental data on mixing in two-layer stratified flows. For different flow regimes (supercritical, subcritical or close to critical) he proposed different equations for predicting the entrainment coefficient based on the bulk Richardson number. Sullivan and List [1994] studied the interfacial mixing in their two-layer stratified flow experiments. They reported that interfacial mixing appears to be dominated by KH instabilities when the gradient Richardson number is less than about 0.4, and it is dominated by shear-driven wave breaking for a larger Richardson number. The rate of entrainment is also related to the gradient Richardson number.

Strang and Fernando [2001] studied turbulent entrainment at the two-layer interface with the moving upper layer and a quiescent, deep dense layer. They found that for the flow with a bulk Richardson number $J < 1.5$ (defined using the depth of the upper layer), the entrainment takes place as if no stratification were present (i.e. independent of J) and the entrainment coefficient is consistent with what is reported by Christodoulou [1986]. For $1.5 < J < 5$, KH billowing is the dominant mixing mechanism and the entrainment coefficient is a function of J. At $J > 5$ KH-type instabilities are suppressed and the entrainment coefficient is reduced by an order of magnitude. Peltier and Caulfield [2003] examined the literature on the mixing efficiency in stratified shear flows. They found that the commonly reported mixing efficiency value of 0.2 can be related to the transition of the flow to turbulence through an intermediate instability of KH type. While caution has to be taken in applying any of the above results, it is clear that interfacial wave activities are directly responsible for interfacial entrainment and mixing. For flow entrainment in downslope flows, Christodoulou [1991] proposed a simple model for evaluating the dilution of a dense 3-D bottom plume, created by the disposal of heavier effluent into a quiescent homogenous ambient. The spreading of the plume is described using the available data, and the dilution was estimated based on matching requirements between the near and intermediate field. Pawlak and Armi [2000] studied the flow entrainment caused by the lower moving layer in arrested wedge flows downstream of the sill crest. They reported that two regimes exist: an initial region of low Richardson number and high entrainment coefficients, and a high Richardson number region characterized by weak entrainment and the collapse of turbulence.

Baines [1975] extended the results of Ellison and Turner [1959] to stratified environments. The dense flow was released at the top of the slope for a finite period of time. The downslope gravity current approaches its level of neutral density where the fluid leaves the proximity of the slope. Rather than flow entrainment, a continuous loss of fluid from the downflow was observed in most cases that he studied. Dallimore and Ishikawa [2001] studied a saline underflow into a lake. The rate of entrainment into the bottom layer is related to the bulk Richardson number (defined with the mean properties of the layer). However, they indicated that the entrainment and mixing is mainly caused by the bottom boundary generated turbulence, which is transported vertically to the density interface. Most of the studies on entrainment in such flows have been dedicated in giving a general relation between the entrainment coefficients and the bulk Richardson number (see Ellison and Turner [1959], Lofquist [1960], Christodoulou [1986], Garcia [1996] and Kessel and Kranenburg [1996]). Dallimore and Ishikawa [2001] studied a saline underflow into a lake relating the rate of entrainment into the bottom layer to the bulk Richardson number (defined with the mean properties of the layer). However, they indicated that the entrainment and mixing is mainly caused by the bottom boundary generated turbulence, which is transported vertically to the density interface.

In downslope flows, however, it is not clear how important is the bottom generated turbulence vs. the interfacial instabilities in causing interfacial entrainment and mixing. Other studies tried to formulate general (empirical) expressions to relate entrainment coefficients in stratified exchange or arrested flows to the Richardson number and the drag coefficient related to the bottom roughness (see for example Hebbert and Loh [1979], Atkinson [1988] or Dallimore and Ishikawa [2001]). A detailed comparison between these models and velocity field data of a shallow underflow into a medium-sized reservoir is given in Fernandez and Imberger [2007]. Morin and Loewen [2004] studied the entrainment caused by the downslope flows in an exchange flow down a sill. They obtained a large entrainment coefficient in the region where K-H waves are generated. Their entrainment coefficient was found to be considerably larger than the predicted values for gravity currents at the same Richardson numbers. In the above studies, both, the thickness of the active layer and the shear layer thickness are used in defining the bulk Richardson number. The entrainment and mixing processes in Strang and Fernando [2001] experiments are controlled by the large eddies with a length scale of the layer depth in the flowing upper layer.

2.4 Barotropic effects

The flow behavior changes when external forces are superimposed on the pure baroclinic flow. A series of papers by Farmer and Armi in the 1980's analyzed the hydraulics of exchange and arrested flows, also through a contraction and a combination of a contraction and a sill, including results with a constant barotropic component. They showed, that maximal two-way exchange with a net barotropic flow requires the presence of two controls, one at the narrowest and shallowest position and another control, whose position depends on the barotropic force: the control must lie on the side from which the barotropic component is directed. In the absence of barotropic flow the two controls coalesce and the interface at the narrowest and shallowest section is half of the total water depth.

Pawlak and Armi [1998], Pawlak and Armi [2000] examined the vortex mechanisms developed in a stratified shear layer subject to spatial acceleration and the subsequent interfacial entrainment and mixing in their physical experiments in downslope currents, including results for flows with a superimposed weak and strong constant barotropic flow. Their outer flow was dictated by a hydraulically controlled wedge flow that provided a spatially accelerating shear layer and baroclinic generation of interfacial vorticity along the density interface above a mild sloping topography. They observed a mechanism of vorticity generation, where the core of the growing vortex was separated from the vorticity source and a second core developed. Using linear stability analysis they observed that one of two modified KH modes was dominant, so that the instability lied in the slower moving stream. Finally, they showed that mixing induced by interfacial instabilities lead to the generation of a sharp density interface near the high momentum stream. Recent studies on unsteady two-layer hydraulics have some relevance to the initial start-up phases of these flows. Baines [1975] (also see his book, Baines [1995]) studied the unsteady flow adjustment of two-layer flow upstream of a bottom sill. Helfrich [1995] examined the effect of the time-dependence (e.g. tidal influence) on exchange flows. Pawlak and Armi [1998] reported the effect of non-uniform velocity profiles. Exchange flows through a barrier are studied by Zhu [2002] and by Cuthbertson and Coates [2004] where a oscillating barrier has been used.

Stommel and Farmer [1953] and later Welander [1984] pointed out that barotropic current fluctuations can influence the state of overmixing, but none of them made a quantitative estimation of the effect. Stigebrandt [1976] developed a simple theory, showing that the two-layer transport capacity of a constriction may be increased considerably by barotropic current fluctuations. This was also confirmed by laboratory experiments. He found out that the barotropic current fluctuations have no effects on the transports for the case $\sqrt{g'H}/(2u_b) \geq 1$, where g' is the reduced gravitational acceleration, H is the total water depth and u_b is the amplitude of the periodic barotropic current. Helfrich [1995] examined the effect of the time-dependence (e.g. tidal influence) on exchange flows.

3 Experimental methods

In this chapter the experimental design (section 3.1) as well as the investigation methods used for the analysis of the interfacial instabilities are described. These are the Particle Image Velocimetry technique (PIV) to obtain velocity distributions in the flow field (section 3.2.1), and Planar Laser induced Fluorescence (PLIF) to obtain information on the concentration in the flow field (section 3.2.2). The experimental errors related to these techniques are highlighted in section 3.3.

The results relative to the experiments are presented separately in chapters 4, 5 and 6.

3.1 Experimental facility

To perform the experiments of a two-layer density-stratified flow over a submerged sill, a 12 m long and 0.6 m wide basin shown in figure 3.1 has been used, which is divided in the middle into two basins, and connected by a channel of reduced width (14.5 cm). The tank has a volume of 6 m^3, which is 10 times bigger than the one used by Morin and Loewen [2004]. It thus, allows a much wider range of flow conditions to be studied, in particular a longer experimental duration with much larger Reynolds numbers. A bottom sill was placed in the channel. To avoid leakages caused by corrosion due to the salt water in the tank, it has been painted using a special paint very resistant to salt water (Sikagard Poolcoat, Otto Chemie, Germany). The sill, first made by concrete, was then fabricated by AKA-Flieg Karlsruhe using foam plastic and it consists of two parts, enabling two different sill slopes which shapes are described by: $z(x) = hcos^2(\pi x/(2L))$, for $-40 < x < 0$, and $z(x) = h - x/5$, for $0 < x < 100$, where $h = 20$ cm and $L = 100$ cm (see figure 3.1). The sill was clamped by two hooks fixed at the bottom.

3.1.1 Experiment execution and set-up optimization

The tank was first filled with fresh water to a desired depth. A plexiglas barrier was then placed at the sill crest to divide the tank in 2 reservoirs. The density difference $\Delta\rho$ was produced by adding salt to the water in the left reservoir. The water temperature was recorded (typically ranging from $12°C$ to $15°C$) and the density of each water body was determined based on the water temperature and salt concentration using the formula $\Delta\rho/\rho \sim 0.7S$ where S is the ratio between the salt weight and the water weight. In the combined PIV/PLIF measurements, probes of the salty water were taken and the density was measured with a density-meter (Anton Paar, DMA 5000, working according to the oscillating U-tube principle) with a resolution in the order of 10^{-6} kg/m^3. To minimize effects of wave reflections at the end of the reservoirs, two grid plates were built at each

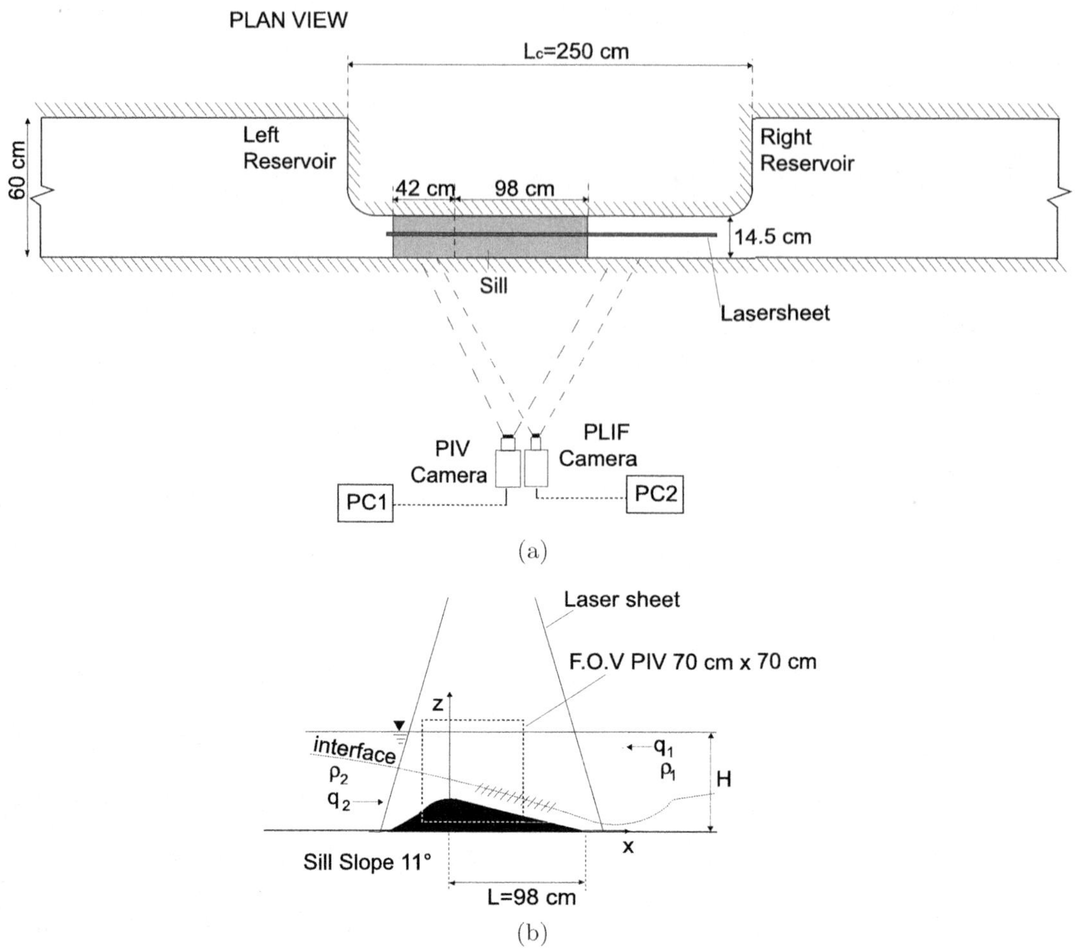

Figure 3.1: (a) Plan view and (b) side view of the experimental set-up.

side end of the reservoirs. For a more detailed discussion about the experimental starting conditions it is referred to chapter 5, section 4.1.

3.1.2 Scales and model similarity

A brief dimensional analysis is performed in order to determine the key parameters influencing the flow behavior in the experimental facility described in the previous section. One of the novelties brought from these series of experiments is a higher Reynolds number (which is defined below), which ranges from 10,000 to 100,000. This is at least one order of magnitude bigger than the exchange/arrested flows investigated before (see for example Pawlak and Armi [1998], Morin and Loewen [2004]). The fluid dynamics of flows at high Reynolds numbers is characterized by the existence of several length and time scales. In the here performed experiments, different parameters play a role on the behavior of the flow development.

The length scale in longitudinal direction L is determined by the development of the shear and mixing layer, and is estimated to be of the order of the right sill length L. The corresponding time-scale is given by $t_a = L/U = L/\sqrt{g'H}$. An acceleration time scale can be defined to measure the local effect of the spatial acceleration on the developing mixed layer, as follows

$$t_a = \left(\frac{c}{\Delta U}\frac{d(\Delta U)}{dx}\right)^{-1}$$

in which c is the propagation speed of the interfacial wave. This time scale will be used to analyze the effect of the spatial acceleration through an acceleration parameter which will be better defined in section 6.5.

The vertical length scale is given by the total water depth H. Moreover, other scales are the buoyant acceleration g', the kinematic viscosity ν, the bottom roughness height k_s and the sill slope, which effect can be represented by the gravitational acceleration g. Finally, the velocity scale ΔU can be expressed as follows:

$$\Delta U = \mathcal{F}(H, g', \nu, k_s, g),$$

where $\mathcal{F}$ denotes some functional dependency. Scaling the velocity scale ΔU with the velocity scale $\sqrt{g'H}$ the following relationship can be obtained:

$$Fr_\Delta = \mathcal{F}_1(Re, k_s/H, g'/g)$$

where $Fr_\Delta = \Delta U/\sqrt{g'H}$ is an overall Froude number and $Re = \sqrt{g'H}H/\nu$ is the Reynolds number. The ratio g'/g represent a comparison between the stabilizing effect of the buoyancy and the destabilizing effect given by the acceleration down the sill. In this study, the density difference between the two layers is small, typically 0.4%, and the Boussinesq approximation can be applied so that this term can be neglected.

By this dimensional analysis, it is evident that the flow depends upon both the Froude and the Reynolds number. The Reynolds numbers reached in these experiments are reduced by a factor of roughly 1000 while the Froude numbers are reduced by a factor 2, as compared to the typical scales reached in a full scale-model (see for example the typical scales for the Strait of Gibraltar in figure 1.3).

As pointed out above, the main flow velocities for hydraulically controlled flows scale with the buoyant acceleration g' and the water depth H (refer to figure 3.1 (b)), so that the velocity scale U is given by $\sqrt{g'H}$. The relevant length scale in vertical direction is given by the water depth H. The corresponding time scale is given by $t = (H/g')^{1/2}$. Considering a ratio between the vertical scales between the model and the prototyp of 1/1000, as typically it occurs in the experiments here, and for the same buoyant acceleration, then timescales and velocity scales in the model are roughly 30 times faster than those at the full scale model.

3.2 Experimental techniques

The velocity and concentration fields were obtained using a Particle Image Velocimetry (PIV) and a Planar Laser induced Fluorescence (PLIF) measurement techniques, respectively. A brief description of the two techniques are given in the following sections.

3.2.1 Particle Image Velocimetry (PIV)

The PIV method is a non-intrusive measurement technique that is capable of measuring two-dimensional velocities at many points in a flow field, simultaneously. The typical experimental set-up for PIV consists of a light source, light sheet optics, seeding particles, a camera, and a PC equipped with a frame grabber and an image acquisition software. Polyamide particles (PA12, Vestosint 2157) with a mean diameter of 200 μm and a density of 1.016 g/cm^3, were added in both reservoirs as tracer material for the velocity measurements with PIV. A 10 Watt Argon-Ion laser (Stabilite 2016, Spectra-Physics lasers) operating in multimode ($\lambda_1 = 488\ nm$, $\lambda_2 = 514\ nm$) has been used as continuous light source. The beam was transmitted through a fiber optic cable to a line generator with spherical lenses (OZ Optics Ltd., Nepean, Ontario). The generated laser sheet at the sill crest had a length of approximately 2 m and a width of 5 mm and was positioned in the middle of the channel. Images of 70 cm x 70 cm were grabbed with a CCD camera (1024 x 1024 Pixels, 8 bits) at a frame rate ranging from 16 to 24 Hz. On the CCD camera, an objective (SIGMA AF EX 1.8/24 DG Macro AF for Nikon) was mounted along with a low-pass filter (DT Cyan, Linos Photonics GmbH, Göttingen) leading to a spatial resolution of 0.0625 $cm/pixel$. The $15,000$ raw images were stored in real time on a raid system. This corresponded to enormous amounts of data with each experiment using approximately 15 Gigabytes of disk space for the raw images and 5 Gigabytes for the vector fields for a total value of approximately 20 Gigabytes of non-processed data for **each** experiment. The raw image pairs were then processed using a PIV algorithm to compute the velocity fields. The basic concept of a PIV scheme is based on the splitting the first image into sub-regions, called interrogation windows, and the second image into sub-regions called search windows. Cross-correlation of these sub-regions within an image pair gives the averaged displacement of all of the particles within the interrogation window. Dividing this displacement by the time separation between the two images gives the instantaneous velocity vector for that interrogation window. This process is then repeated for all the interrogation regions to obtain the velocity field over the entire field of view. With the software package DaVis (LaVision) the velocity fields were computed using a cross-correlation PIV algorithm. For this purpose an adaptive multipass routine was used, starting with an interrogation window of 32 x 32 pixels and a final window size of 16 x 16 pixels with 50% overlap. Each vector of the resulting vector field represents an area of 0.6 cm x 0.6 cm. The velocity vectors were post-processed using a local median filter.

3.2.2 Planar Laser induced Fluorescence (PLIF)

The use of LIF for measuring concentration fields is attractive because it is non-intrusive and can measure scalar concentrations and their fluctuations at multiple points simultaneously. A laser beam illuminates a dye tracer and fluorescent light is emitted. To translate dye concentrations into salt concentration it is important that the two solutes have the same diffusivity (also known as diffusivity matching). The Schmidt (Sc) number is defined as the ratio between the kinematic viscosity ν and the mass transfer diffusion coefficient κ_c. Rhodamine 6G has a Schmidt number ranging from 600-1200, depending on the reference. The Schmidt number for salt is approximately 700 thus the use

Rhodamine 6G with salt is a good combination for diffusivity matching (see for example Troy and Koseff [2005] and many others). In all the concentration measurements performed, Rhodamine 6G was used as dye and was added to the salt water. In general, the fluorescence is the consequence of the decay of a molecule from an excited to a fundamental state, involving spontaneous emission. The exited state can be reached by the absorption of laser radiation (for Rhodamine 6G at 520 nm) whose wavelength is within the absorption range of the molecule spectrum. The Rhodamine 6G dye emits light at higher wavelengths than the laser and this is very helpful in separating the fluorescent light from the incident laser radiation. The emission of Rhodamine 6G is at a wavelength of 590 nm. An orange high-pass filter (DT Orange, Hama GmbH, Monheim) with a cutoff of 550 nm was placed on the camera to filter out the reflections at lower wavelengths and so, to reduce the noise in the concentration measurements and to enable later combined PIV/PLIF measurements. Images of 70 cm x 70 cm were grabbed with a CCD camera (1280 x 1024 Pixels) at a frame rate of 210 Hz. On the CCD camera, an objective (SIGMA AF EX 1.8/24 DG Macro AF for Nikon) was mounted, leading to a spatial resolution of 0.0625x0.0625 $cm/pixel$. The $15,000$ raw images were stored in real time on a raid system. This corresponded again to enormous amounts of data with each experiment using approximately 20 Gigabytes of disk space of non-processed data.

3.2.3 Calibration procedure for PLIF

If the intensity of the laser sheet is known in every position the concentration of the dye can be determined as the relation of the intensity is directly proportional to the concentration of the fluorescent dye

$$S = \alpha C_{rho} I \tag{3.1}$$

where S is the fluorescent signal, C_{rho} is the concentration of Rhodamine 6G and α is some calibration constant. In the literature numerous parameters have been reported which can affect the quality of the experimental results using Laser induced Fluorescence but only attenuation was believed to be a significant problem. The diffusivity matching could be neglected since Rhodamine 6G and salt have the same Schmidt number and the refracting index matching was also neglected as long as the salt concentration was very low. As the incident laser beam travels through the water attenuation of its energy takes place. Concentration of solutes increase this effect and were found to be additive. Koochesfahani and Dimotakis [1985] found for the decay of laser intensity along its path the exponential function $I = I_0 e^{a(z-z_0)}$ where I is the light intensity at location z, I_0 is the light intensity at the location z_0 (i.e. at the water surface) and a is the attenuation coefficient. The attenuation due to different solutes are additive and also the attenuation of the fresh water has to be taken into account. Daviero and Maile [2001] expressed the attenuation a as the sum of the attenuation of the single solutes. The total attenuation is given by: $a = a_w + a_{rho} + a_{salt}$, where a_w, a_{rho} and a_{salt} are respectively the attenuation due to fresh water, Rhodamine 6G and salt. Daviero and Maile [2001] found for the fresh water attenuation a value between 0.0011 cm^{-1} and 0.0045 cm^{-1}. The attenuation due to Rhodamine 6G and salt are dependent on the concentration and are given by: $a_{rho} = \varepsilon_{rho} C_{rho}$ and $a_{salt} = \varepsilon_{salt} C_{salt}$, where C is the concentration of the solute and values

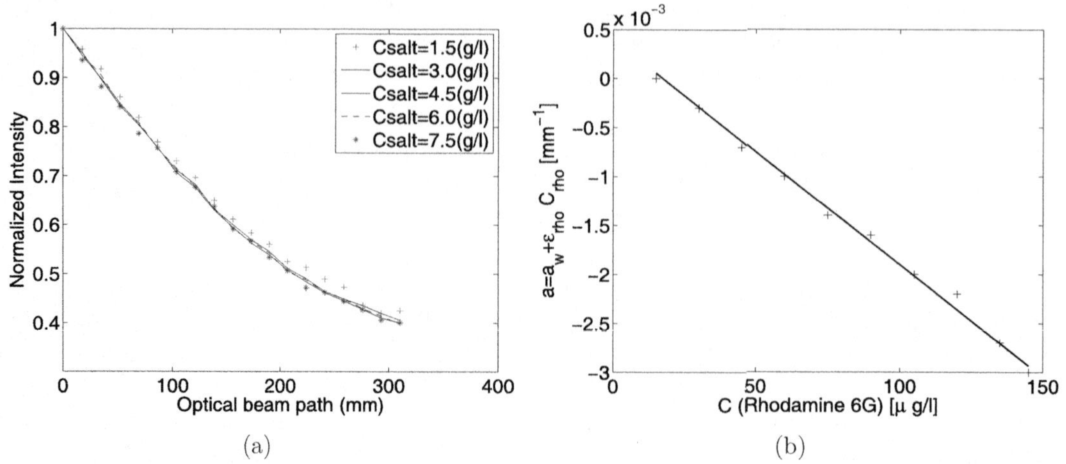

Figure 3.2: a) Intensity decrease in vertical direction along a the optical beam path due to water and salt. The intensity does not change increasing the salt concentration; b) Total attenuation coefficient $a = a_w + \epsilon_{rho} C_{rho}$, with $a_w = 0.0004$ and $\epsilon_{rho} = -2.3 * 10^{-5}$ for different concentrations of Rhodamine 6G.

of $2.30 \cdot 10^{-4}\ cm^{-1}(\mu g/l)^{-1}$ and $1.24 \cdot 10^{-4} cm^{-1}(g/l)^{-1}$ for the coefficient ε were reported, for Rhodamine 6G and salt, respectively. Important to note is that for Rhodamine 6G the concentrations are given in $\mu g/l$ and for salt in g/l. Thus, the attenuation due to salt can be neglected for small concentrations, as also shown in figure 3.2 (a), and the total attenuation coefficient can be rewritten as

$$a = a_w + \varepsilon_{rho} C_{rho} \tag{3.2}$$

A series of calibration experiments have been performed in order to determine the attenuation coefficients and to compare them with the values in the literature. The experiments have been conducted in a 250 cm long, 10 cm deep and 50 cm high plexiglas tank. This was filled to a 40 cm depth. The calibration experiments showed for the attenuation coefficient of Rhodamine 6G (a_{rho}) and of water (a_w) a very good match (see figure 3.2 (b)) with the coefficients found in literature (cf. Daviero and Maile [2001]).

The influence of the exposure time was examined to determine its influence on the noise ratio and on recorded intensity. The longer the exposure time, the higher the intensity; however, the exposure time is limited by the recording frequency of the camera (10 Hz) and the storage speed capacity of the PC. Figure 3.3 show intensity profiles at a horizontal section for different exposure times: the influence of the exposure time is significant, while the influence on the noise is negligible. Thus, a exposure time of 20 ms was chosen, which is the maximal value which can be supported by the used CCD camera.

The major problems with the PLIF calibration was a non uniform and a non stable light distribution. Power fluctuations were observed not only from one day to the other, but also during the same experiment, when the laser was not switched off. Figure 3.4 show this behaviour of the laser light distribution. The causes of these power fluctuations are due to the optical set-up: the laser is operating in multimode (there is no prisma to

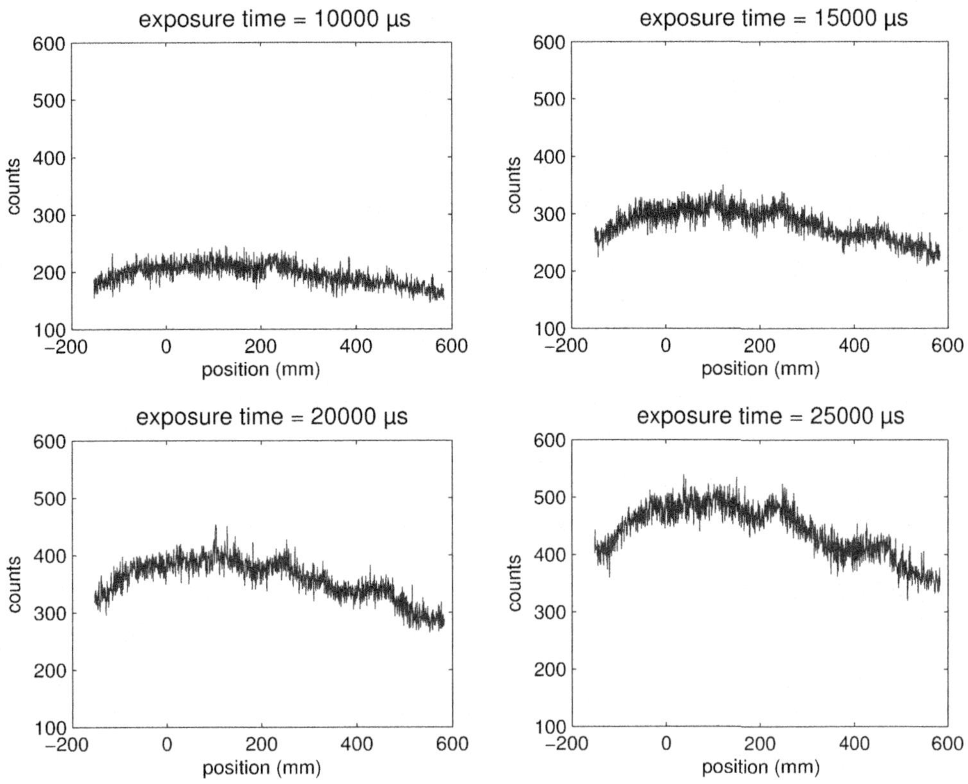

Figure 3.3: Horizontal intensity profiles for four different exposure times.

separate the two wavelengths), so that both wavelengths of the laser light are leaded to the optical fiber glass cable. First, the beams reflect different in the optical fiber glass cable and second, they can interact.

This problem of an non-uniform and non-constant light distribution made it impossible to calibrate the laser sheet before performing an experiment. Instead, a simple calibration procedure is proposed in the following section.

3.2.4 Final calibration procedure

When examining the development of the flow down the sill slope, three zones can be distinguished, in the following called 1, 2 and 3 (see figure 3.5). The first zone (zone 1, see figure 3.5) includes the fresh water region, where the laser light can travel with relatively low attenuation. The third zone (zone 3, see figure 3.5) is characterized with the highest concentrations and the effects of attenuation are very strong. In the second zone (zone 2, see figure 3.5), the concentration can vary from the lower to the highest value randomly. With this schematization of the flow, the calibration procedure proposed here, is based on a series of assumptions and simplifications:

1. every pixel column in the picture, is divided by its maximum, so that the concentration is ranging from 0 to 1. This is the only way to calibrate the non uniformly

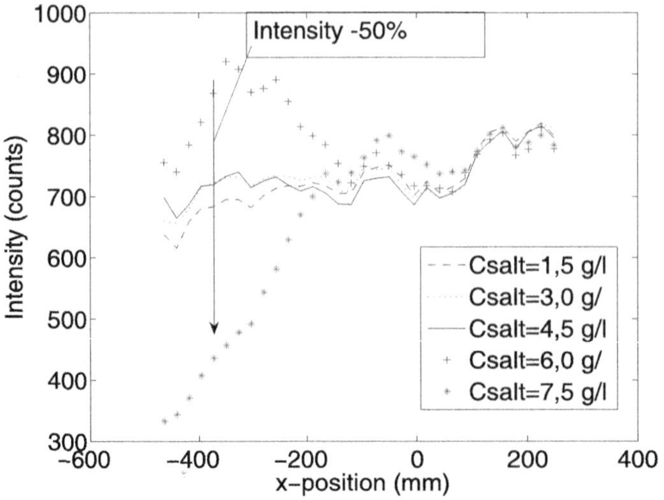

Figure 3.4: Laser intensity distribution along three horizontal profiles at the water surface during the same experiment of 20 minutes duration.

distributed laser light, at every time. Zone 1 is then defined as the zone where the concentration is ranging from 0% to 0.5% while zone 3 is defined as the zone where the concentration is ranging from 99.5% to 100%. In zone 2, which is the region of interest, the relation between the pixel intensities and concentration is assumed to be *linear* and thus, attenuation effects are neglected in this zone. However, this

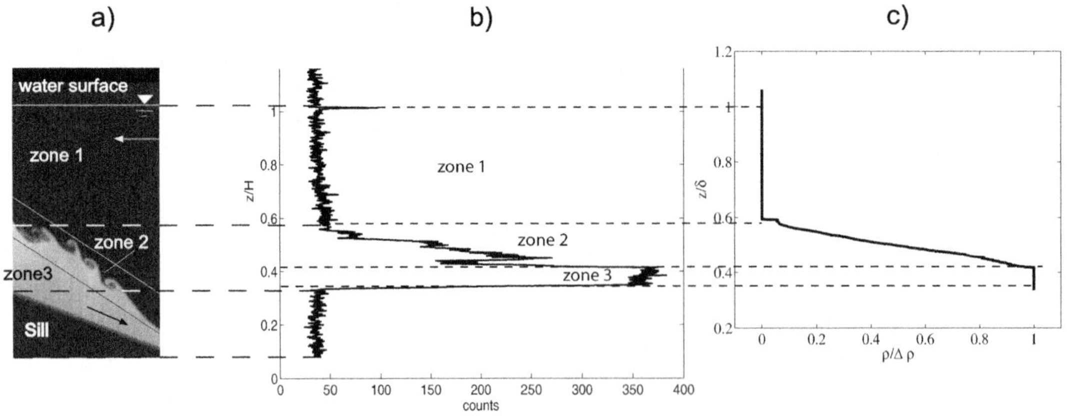

Figure 3.5: Definition sketch for the scheme used in the calibration procedure. a) Section of an instantaneous picture, b) raw intensity profile at the left boundary of the instantaneous picture and c) averaged and corrected density profile after calibration. In zone 1 there is only fresh water flowing from the right to the left; zone 3 is the denser layer characterized with the highest (known) concentration and flowing from the left to the right down the sill; zone 2 is the intermediate layer where mixing and entrainment occur; on the right the vertical intensity profile of the instantaneous image is shown.

assumption reproduce well the real concentration distribution, as long as the noise in the vertical density profile is small (see figure 3.5) and as the thickness of the zone 2 is not too large. It is estimated to be 25% of the total water depth;

2. with the calibration procedure described above, it is intrinsically assumed that the concentration in zone 3 in not varying in longitudinal direction down the sill. However, the eventual error is estimated to be below the precision enabled by the experimental set-up. The cases where this assumption will produce higher errors are the experiments with bottom roughness;

3. the beam paths are assumed to be vertical and not radial.

The following section also deals with the error estimation due to these assumptions.

3.3 Experimental errors

- **PIV:** the velocity measurements are subject to error due to the velocity gradients, particle density, particle diameter, out-of-plane-motion and peak locking [Cowen and Monismith, 1997]. The chosen PIV particles had a size between $100 \div 300\ \mu m$, giving a non-dimensional particle diameter of 0.2 $pixels/pixel$. Particles smaller than one pixel always occupy one pixel in a PIV image; thus, the true position of the particle cannot be resolved within a pixel. A small ratio of particle size to pixel size was inevitable in this experimental set-up due to the large field of view required for the study. Peak locking refers to the bias that occurs when the estimated location of the correlated peak is shifted towards the next integer value. La Vision software delivers that 2% of the streamwise velocity vectors and 18% of the vertical velocity vectors were affected by peak locking, to give a mean value of 10% for the total velocity. The averaged Q-factor is 2.4. The surprisingly good density probability distribution and the value of the Q-factor can be explained by the fact that the La Vision software identifies a group of particles as one single particle. However, there is no possibility of determining the percentage of the particles affected by this grouping effect of the software and it does not matter, since we are interested in what happens at larger scales.

  The mean values of the largest velocity gradients in the streamwise and vertical directions were computed from the raw PIV data and evaluated to be approximately 2% (cf. Morin and Loewen [2004]). In figure 5 (e) of Cowen and Monismith [1997] the appropriate errors due to the velocity gradients are presented. The total error due to a 2% gradient of a particle diameter of 3 pixels is estimated to be 0.08 pixels (see figure 5 (a) in Cowen and Monismith [1997]). Out of plane motion has been neglected, as the flow can be treated as two-dimensional. The total error can be then estimated as the error of the particle size and velocity gradients to 0.08 pixels, which corresponds to a value, in the worst case, of $(spatial\ true\ resolution) \times (error) \times (acquisition\ frequency) = 0.0625\ cm/pixel \cdot 0.08\ pixel \cdot 24\ Hz = 0.12\ cm/s$, expressed in velocity units. Given the velocities encountered in the experiments, the experimental error in the instantaneous velocity is estimated to be approximately 3%.

- **PLIF:** the errors due to the first assumption described in section 3.2.4 (due to the energy loss in zone 2) are estimated to be approximately 8%-10%, using the attenuation coefficients ($a_w = 0.0028\ cm^{-1}, a_{rho} = 0.00023\ cm^{-1}(\mu g/l)^{-1}$) found during the calibration experiments. The error due to the non uniform distribution of the laser sheet is not so easily calculated. The calibration algorithm takes the maximum value in a vertical column of a picture which corresponds to different initial laser intensities. This error is increasing with increasing values of the diverging angle of the beam paths. Therefore, the biggest errors will occur at the sides of the picture and more on the left side. A more precise estimation of this error cannot be made unless the distribution of the laser light is known.

Results relative to the physical experiments are presented in chapters 4, 5 and 6.

4 Unsteady start-up conditions and barotropic responses

In general, the flow processes involved in a two-layer density-stratified system are an interaction of purely baroclinic effects (due to density differences) and the associated interfacial instabilities, on one hand, and of additional barotropic components (due to various types of steady or unsteady external forcing), on the other hand. In this study, the baroclinic component of the flow is obtained using different salt concentrations, while the pulsating external forcing (i.e. the barotropic component) is inevitably produced due to the starting conditions related to the experimental facility. The combined effects of baroclinic and unsteady barotropic components are ubiquitous in geophysical and environmental flows, as for example in sea straits, where metereological or astronomical tides are superposed to the baroclinic exchange flow or in exchange flows between two confined reservoirs (like harbor oscillations) or different portion of a lake, where basin oscillations often occur for example due to wind effects.

In this chapter, experimental results relative to the barotropic oscillation superposed to the baroclinic flow are presented. The chapter is organized as follows.

The different starting conditions for the experiments are discussed in section 4.1. The theoretical approach predicting the flow rate oscillations and the related period is presented in section 4.2. The experimental program is presented in section 4.3. In section 4.4 the baroclinic flow with superimposed barotropic oscillation is described showing that significant effects on the exchange flow are produced by the barotropic component.

Further effects of the barotropic oscillation on the flow development and on interfacial wave characteristics and entrainment coefficients under conditions of higher Reynolds numbers are described in the next chapter 5.

4.1 Different starting conditions

There are different possibilities for starting the experiments given the experimental facility described in the previous section. A first possibility, which has also been used in the majority of previous studies in stratified exchange flows, is to fill the tank to a desired water depth. After inserting a separating barrier to divide the tank in two reservoirs, a known amount of salt is added in one of the two reservoir to generate the density difference. This configuration for the starting condition is illustrated in figure 4.1 (a). Referring to figure 4.1 (a), if $\Delta h = 0$ as assumed for this first possibility, the density difference between the two reservoirs produces a net hydrostatic pressure force on the separating wall prior to the experiment beginning, i. e. $F_s > F_w$, in which F_s is the force on the partition from the salt water side, and F_w on the fresh water side. This initial global pressure unbalanced state generates a net flow through the flume in longitudinal

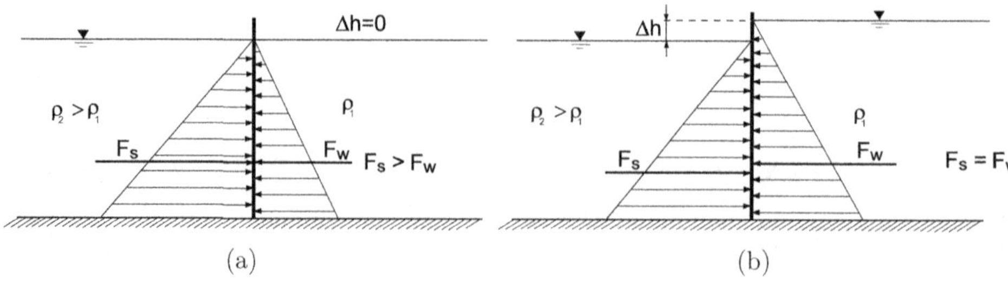

Figure 4.1: Sketch of different experimental starting conditions. (a) Hydrostatic pressure resultant force is unequal but the water depths are at the same level; (b) Global balance equilibrium, but different water depths.

direction and thus an impulsive barotropic forcing. This results finally in an oscillation of the free surface, which will be superimposed onto the baroclinic exchange flow once the wall is removed, leading to wave like responses.

An other possibility to start the experiments is to add the correspondent amount of volume of water in the fresh water reservoir $\Delta h = h_2(\sqrt{\rho_2/\rho_1} - 1)$, with h_2 denoting the water depth in the reservoir filled with salt water and ρ_2 and ρ_1 being the density of the salt and fresh water, respectively. Thereby, a global balance equilibrium prior to the experiment beginning is established, i.e. $F_s = F_w$ (cf. figure 4.1 (b)). In this second configuration the unsteady barotropic component will be generated due to the depth difference Δh at the beginning of the experiment: surface a surge-wave will be created once the barrier is removed.

In the experiments, different methods were tried covering the above two extremes, including slow removal of the barrier. There are different ways to damp the barotropic oscillation, however, in all cases, the experimental evidence shows that there is no way to completely avoid the superposition of additional external forces on the baroclinic flow in experiments on two-layer exchange flows, given this experimental facility. Thus careful attention should be paid in applying the different experimental starting conditions and their effects on the flow development should be carefully considered.

In the following section a simplified model adapted for our experimental conditions is developed in order to deliver a general relation which can predict the flow rate oscillations and the related period, in an experimental two-basins set-up.

4.2 Simplified model of barotropic oscillations superimposed on exchange flows

In this section a simplified model to study barotropically induced oscillations is derived, in an experimental two-basins set-up. The superimposed barotropic flow which is inevitably generated at the beginning of the experiment is also known as Helmholtz oscillation. The Helmholtz oscillation is characterized by a periodic mass exchange through a narrow strait and associated spatially uniform changes in the basins, representing a simple kind of resonance produced by an oscillatory external forcing. It would be useful to predict

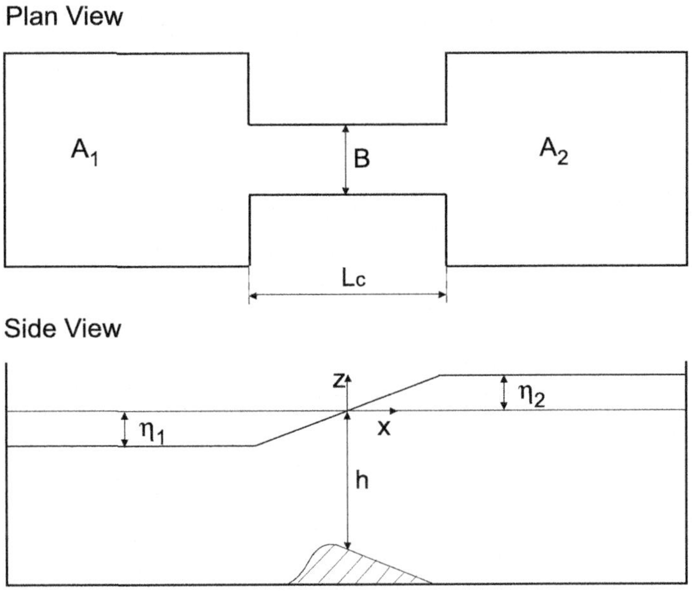

Figure 4.2: Sketch of the hydraulic system used for the model in section 4.2.

the oscillation periods and the flow rates, given the experimental conditions. For this purpose the oscillating flow is considered as an unsteady, one-dimensional flow between the two basins as shown in figure 4.2. Neglecting in first order the viscous term, the momentum equation for the horizontal velocity component u is

$$\frac{\partial u}{\partial t} + u\frac{\partial u}{\partial x} = -\frac{1}{\rho}\frac{\partial p}{\partial x}. \tag{4.1}$$

Assuming that the volume of the basins are much larger than that of the channel, the two basins are related by

$$A_1\eta_1 = -A_2\eta_2, \tag{4.2}$$

where A_1 and A_2 are the basin areas and η_1 and η_2 are the free surface displacements with respect to the still water elevation. The continuity equation reads

$$A_2\frac{\partial \eta_2}{\partial t} = qB, \tag{4.3}$$

where q is the flow rate per unit width and B is the width of the channel, assumed constant. Assuming a small surface deflection, η_1 and $\eta_2 \ll h$, the momentum equation (4.1) becomes

$$\frac{1}{h}\frac{\partial q}{\partial t} + \frac{q}{h}\frac{\partial (q/h)}{\partial x} = -\frac{1}{\rho}\frac{\partial p}{\partial x}. \tag{4.4}$$

The pressure gradient is represented by a linear free surface deflection

$$\frac{\partial p}{\partial x} = \frac{(\eta_2 - \eta_1)\rho g}{L_C} = \frac{\eta_2(1 + A_2/A_1)\rho g}{L_C}, \tag{4.5}$$

where L_C is the length of the connecting channel. Equation (4.4) becomes

$$\frac{1}{h}\frac{\partial q}{\partial t}+\frac{q}{h}\frac{\partial(q/h)}{\partial x}=-\frac{g\eta_2(1+A_2/A_1)}{L_C}. \tag{4.6}$$

Differentiating equation (4.6) with respect to time, we finally obtain

$$\frac{1}{h}\left[\frac{\partial^2 q}{\partial t^2}-\left(\frac{2}{h^2}\frac{dh}{dx}\right)q\frac{\partial q}{\partial t}\right]=-\frac{gqB(1+A_2/A_1)}{L_C A_2}. \tag{4.7}$$

At the sill location, $\frac{dh}{dx}=0$, equation (4.7) reduces to the wave equation

$$\frac{\partial^2 q}{\partial t^2}+\left[\frac{ghB(1+A_2/A_1)}{L_C A_2}\right]q=0. \tag{4.8}$$

The general solution is then given by:

$$q(t)=Asin(\omega t+\phi), \tag{4.9}$$

where $\omega=\sqrt{k}$, $k=ghB(1+A_2/A_1)/(L_C A_2)$, and ϕ is the initial displacement, which can be set to zero with the initial condition $q|_{t=0}=0$. The associated period is given by:

$$T=\frac{2\pi}{\omega}=2\pi\sqrt{\frac{L_C}{g(1+A_2/A_1)}\frac{A_2}{Bh}}. \tag{4.10}$$

Equation 4.10 can be rearranged to give

$$T=\frac{2\pi L_C}{\sqrt{ghA_C\frac{A_1+A_2}{A_1 A_2}}}, \tag{4.11}$$

where $A_C=L_C B$ is the surface area of the channel. If $A_1 \gg A_2$ (or $A_2 \gg A_1$), like the connection between the ocean and a harbor, the above equation can be further simplified. Equation (4.11) is derived assuming that the water level in both tanks rises and falls in unison. In this study, the disturbance starts from the channel, and it travels through the tank before it is bounced back. This should suggest that the period should be increased by a factor which accounts for the time spent by the gravity wave to reach the end of the channel. This correction could be made by using for example a larger length of the channel and reduced areas for the reservoirs. On the other hand, the velocity in the channel, which is proportional to $\sqrt{gh}$, is higher than assumed in the model, as we assume the smallest water depth h at the sill crest, and this might compensate the assumption of unison rise and fall of the water levels in the reservoirs. The very good agreement of the observed periods with the periods predicted by equation (4.11) presented in the next section will confirm this.

Finally, we expect that in addition to the above frequency other higher frequencies will be present due to the limited dimensions of the tank (side wall effects), which can be calculated by the classical relations valid for harbor hydrodynamics theory (see for example Dean and Dalrymple [1984]) as well as lower frequencies related to internal seiches.

In the following sections and in the next chapter it is shown that due to the starting conditions described above and to the resulting unsteady barotropic forcing, significant effects are generated on the exchange flow.

Exp.	g' $[cm/s^2]$	H $[cm]$	f_{aq} $[Hz]$	Starting condition (Fig. 4.1)	Re_* $\times 10^4$	Re $\times 10^3$	Fr -	J -
1	3.09	40.5	16	a)	4.5	3	1.02	0.2
2	3.09	51.0	20	b)	6.3	8	1.04	0.29
3	3.09	59.7	20	b)	8.2	10	0.82	0.5
4	5.81	41.0	10	b)	6.0	4	0.97	0.21
5	5.81	61.0	16	a)	11.0	12	0.99	0.32
6	0	40+2	24	b)	-	-	-	-
7	0	40+6	24	b)	-	-	-	-

Table 4.1: PIV-measurement program with the notations used: g' is the reduced gravity and H is the total water depth. In all experiments a height of the sill crest of $h_c = 20cm$ was used. The sill slope was in all the experiments 0.2 and smooth. Experiments 6 and 7 have been performed using only fresh water. f_{aq} represents the acquisition frequency. Finally, the possibilities illustrated in figure 4.1 for the starting conditions of the experiments are reported.

4.3 Experimental program

A listing of all the performed experiments with their conditions is given in table 4.1. Experiment 1 was performed with a water depth of $H = 40\ cm$ and a buoyant acceleration of $g' = 3.09\ cm/s^2$. Images were captured with an acquisition frequency $f_{aq} = 16\ Hz$. In experiments 2 and 3 the water depth was increased to $H = 50\ cm$ and $H = 60\ cm$, respectively, giving thus higher Reynolds numbers compared to that of experiment 1. Hereby the acquisition frequency was 20 Hz for both experiments. Experiment 4 was performed with a water depth of $H = 40\ cm$ and a buoyant acceleration of $g' = 5.81\ cm/s^2$, resulting in a Reynolds number $Re = 4,000$ using an acquisition frequency of 10 Hz. In experiment 5, with the highest value of the Reynolds number, the water depth was increased to $H = 60\ cm$, with the same buoyancy acceleration as in experiment 4 and the same acquisition frequency as in experiment 1. Experiments 6 and 7 were performed using only fresh water, with the water depths being different prior to the experiment beginning, with a depth difference of $\Delta h = 2\ cm$ and $\Delta h = 6\ cm$, respectively.

Referring to table 4.1, experiments 1 and 5 were conducted using the first starting possibility discussed in section 4.1 (figure 4.1 (a)), starting the experiment with the water depths in the left and right reservoirs at the same level. Experiments 2, 3 and 4 were conducted adding the correspondent amount of fresh water in the fresh water reservoir, thus using the second starting possibility described in section 4.1 (figure 4.1 (b)).

4.4 Barotropic oscillation in the exchange flow

In the following, the vertical lengths are non-dimensionalized with the total water depth H while the horizontal lengths are non-dimensionalized with the length of the sill on the right hand side of the sill crest L (see figure 3.1 (b)) and the channel width B. The origin of the longitudinal coordinate is posed at the sill crest. Moreover, the velocity scale is given by $\sqrt{g'H}$ and the per unit-width flow rate scale by $H\sqrt{g'H}$.

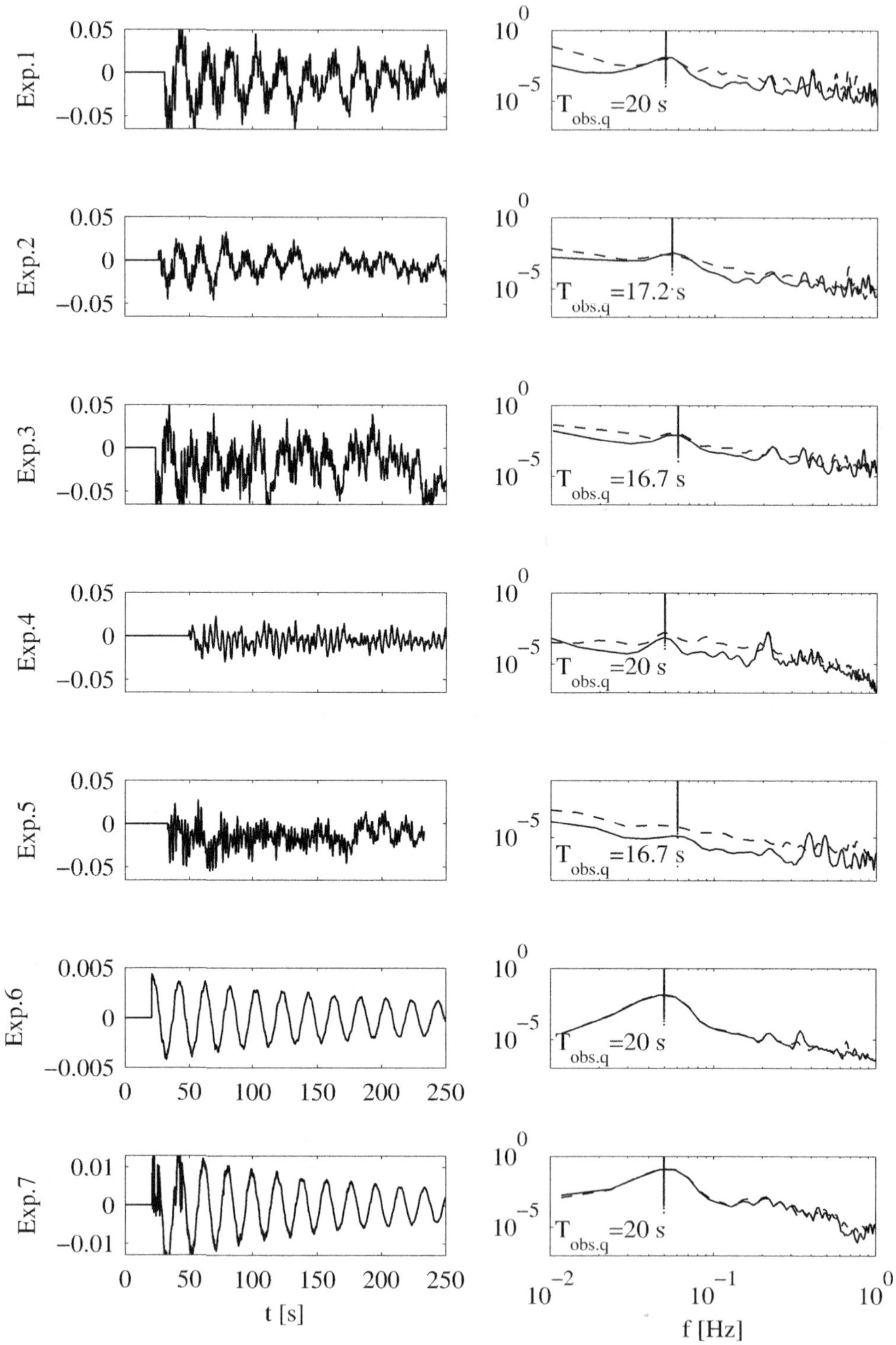

Figure 4.3: Normalized net flow rate per unit width at the sill crest and at $x/L = 0.4$ plotted over time (left column) with relative spectral distribution (right column) for all the experiments. Continuous lines in the right column give the spectral distribution at the sill crest, the dashed lines at $x/L = 0.4$. In experiments 6 and 7 the net flow rate is given in m^2/s.

For the data processing, a number of images of 5,000 was used. In figure 4.3, left column, the normalized total flow rate approximately at the sill crest per unit width is plotted over time for all the performed experiments. In experiments 6 and 7 the net flow rate is normalized with the gravitational acceleration g. It can be noted that for all the performed experiments the flow rate is oscillating in time around a zero value, thus indicating a net flow oscillating periodically from one side to the other side, independent of the chosen method for the experimental starting conditions. The amplitude of these oscillations is decaying in time due to frictional effects. In the right columns of figure 4.3 the spectral distribution of the net flow rates at the sill crest are shown, with dominant periods $T_{obs.q}$. From figure 4.3 it appears that these frequencies are caused by the influence of the pulsating barotropic component, because they are equal for each experiment with the same water depth H and do not depend on the buoyant acceleration. This is additionally proven by considering figure 4.4 (a) which gives the time-shifted velocity profiles at the same location for experiment 1: baroclinic effects would result in a change in the area of the velocity profiles between upper and lower layer, while barotropic effects would cause a translation in longitudinal direction of the entire profile in time. Figure 4.4 (a) shows that the profiles translate in the streamwise direction.

The results presented in figure 4.3 also demonstrates that the second possibility discussed in section 4.1 does not really solve the problem of pulsating barotropic forcing generation, as the same oscillations in the flow rate time sequence and the same peaks in their spectral distributions as in experiments 1, 4, and 5 can be observed.

Spectral analysis was also conducted on the time series of the shear interface, here defined as $(h|_{\overline{u}=0.85} + h|_{\overline{u}=0.15})/2$, where $\overline{u}$ is the normalized velocity ranging from 0 to 1. At the sill crest, low frequencies could be observed and the corresponding periods ($T_{obs,i}$) are also resumed in table 4.2 along with the periods observed also by Morin and Loewen [2004]. From table 4.2 it can be noted that the periods of oscillation of the interface are slightly different from those of the net flow rate. This difference in the periods of the oscillation of the interface and flow rate has been observed previously by Morin and Loewen [2004] (cf. table 4.2), and can be explained as follows: the oscillation in the net

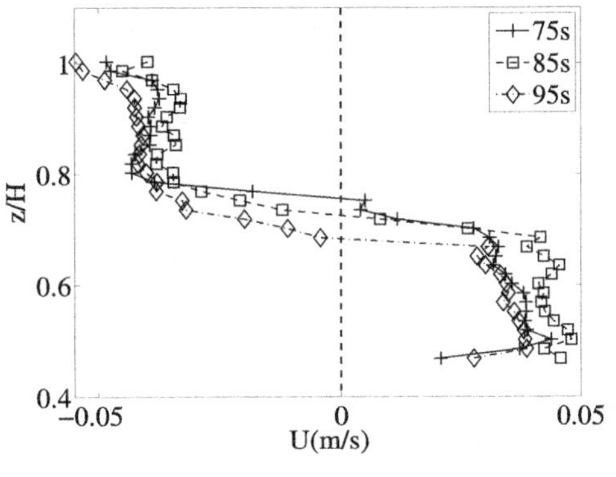

Figure 4.4: Velocity profiles at the sill crest for experiment 1. The time steps are 10 s. The profiles translate in the longitudinal direction thus suggesting that barotropic forces are responsible for the flow rate oscillations.

	$H-h$ [cm]	$T_{obs.i}$ [s]	$T_{obs.q}$ [s]	$T_{Eq.(3.11)}$ [s]	$T_{obs.q}-T_{obs.i}$ %
Morin et al (2004)	20	14–24	14–17	14.5	15.4
This study	20	16.7	20.0	22.5	16.5
	30	14.3	17.2	18.3	16.8
	40	13.3	16.7	16.2	20.3

Table 4.2: Comparison between the observed periods in the spectral distribution of the experimental net flow rate ($T_{obs.q}$) for this study and the study of Morin and Loewen (2004) and the predictions by the theoretical model presented here for the experimental data ($T_{Eq.(3.11)}$). The experimental periods of the flow rate are also compared to those observed in the spectral distribution of the interface position ($T_{obs.i}$) showing discrepancies ($T_{obs.q}-T_{obs.i}$). h represents the height of the sill crest.

flow rate is controlled by the translation motion of the fluid through the channel. The fluctuations of the interface, however, are the result of two mechanisms, namely surge-like accumulations at the sill crest triggered by the barotropic flow and the additional baroclinic wave growth at the interface (i.e. even for steady exchange flow). As can be seen in table 4.2, apparent non-linear interaction between these mechanisms causes a shorter interfacial wave period $T_{obs,i}$ as opposed to the pure barotropic flow fluctuation period $T_{obs,q}$.

The flow rate driven by the barotropic forcing due to the initial experimental conditions can be computed and compared with the measurements. The corresponding Δh due to the density difference is 0.18 cm and 0.12 cm for experiments 1 and 3, respectively. The volume difference between the two tanks is $\Delta h A_1 = 3.6\ l$ and 5.4 l, respectively. This volume difference can drive the flow rates $q = 0.36\ l/s$ and $q = 0.6\ l/s$, respectively, if the volume is divided by half of the barotropic oscillation period $T_{obs,q}$ (cf. figure 4.3, right column). The normalized flow rates ($q/\sqrt{g'H}HB$) give 0.055 and 0.051 respectively, which can be well compared with the flow rate amplitudes plotted in figure 4.3, left column.

In experiments 6 and 7 both basins were filled with fresh water and with a depth difference Δh of 2 cm and 6 cm, respectively, across the barrier between the two basins. The time series of the net flow rate per unit width at the sill crest and the spectral distributions for experiment 6 and 7 are presented in figure 4.3. Again, the same peaks for these two experiments can be recognized as compared to the other experiments with a density difference. The volume difference between the two tanks is $\Delta h A_1 = 60\ l$ and 180 l, respectively, which can drive normalized flow rates $q/(Bg) = 0.0042$ and 0.013 for experiment 6 and 7, respectively. These values are again well comparable to the amplitudes plotted in figure 4.3. Figure 4.3 for experiments 6 and 7 also shows that the depth difference Δh does not influence the frequency, but only the amplitude of the oscillation. It is also worth noting that while in the experiment 6 and 7 the mean value of the flow rate is zero, while in the other experiments the mean value seems to be negative, i.e. more fresh water is flowing from the right to the left as salt water from the left to the right. This is probably due to three dimensional effects and to the difficulty in resolving the lower layer velocity.

A comparison between the periods observed in the flow rate oscillation in the experiments performed in this study, and the periods observed by Morin and Loewen

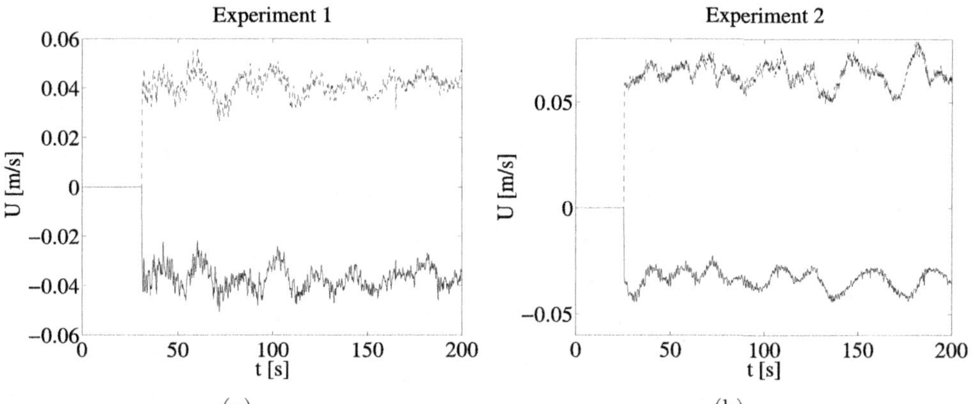

(a) (b)

Figure 4.5: Instantaneous horizontal velocities in the upper (continuous line) and lower (dashed line) layer at the instantaneous velocity interface position for experiment 1 (a) and experiment 2 (b). The time series of the velocities in the upper and lower layer are in phase, thus showing that the oscillations are caused by barotropic forces rather than by internal seiches, in which case they would be out of phase.

[2004], with the periods given by equation (4.11) is given in table 4.2 and it shows excellent agreement. In our experimental facility, the sill height h_s is given by 0.2 m, $A_1 = A_2 \approx (0.6 \cdot 5)\ m^2 = 3\ m^2$, $B = 0.145\ m$ and $L_C = 2.5\ m$.

Another proof to show that the oscillations in the flow rate are caused by barotropic forces rather than by higher modes of internal seiches is given in the figures 4.5 and 4.6. In figure 4.5 the time series of the instantaneous horizontal velocities at the velocity interface position in the lower (continuous lines) and upper (dashed line) layer are plotted for experiment 1 (a) and 2 (b). The time series of the velocities in the upper and lower

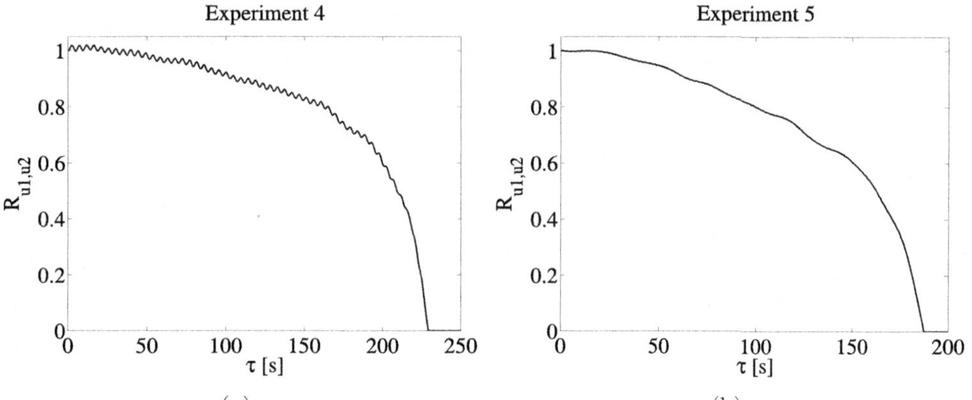

(a) (b)

Figure 4.6: Correlation of the velocities in the upper and lower layer for experiment 4 (a) and experiment 5 (b). The correlation functions remains at a value of nearly one for a long time, thus demonstrating that the velocities are in phase. This suggests again that barotropic forces are responsible for the frequencies resumed in table 4.1.

layer are in phase: for internal seiches they would be out of phase. Moreover, figure 4.6 shows the time-averaged spatial correlation of the velocities in the upper and lower layer ($u_i = \int u_i^2 dz / \int u_i dz$) showing to remain at a value of unity for a long time. It is of interest to compare the above frequencies with that of basin internal seiches. As described by Morin and Loewen [2004], an approximate estimate of the seiche period can be obtained assuming the reservoir is a closed two-dimensional basin with a length $L_{tot} = 12\ m$. The propagation speed of the internal interfacial wave is given by $c = \sqrt{g' z_1 z_2 / (z_1 + z_2)}$, where z_1 and z_2 are the water depths of the lower and upper layer, respectively. The fundamental mode of the internal seiche has thus a period of $T_s = 2L_{tot}/c$. Assuming the interface is at the middle depth of the reservoir, i. e., $z_1 = z_2$, the periods of the internal seiche are given by $T_s = 610\ s$, $498\ s$, $432\ s$, $445\ s$, $315\ s$ for experiments 1, 2, 3, 4, 5, respectively. The peaks relative to these periods can not be observed in figure 4.3 because the steady period duration of the experiments is similar to that of the internal seiche, so that not enough data for the spectral analysis is available.

Finally, it is observed that as an additional effect the superposed barotropic oscillation influences the properties of the flow regimes: generally, for purely baroclinic flows, the maximal exchange regime is steady, but in the case of superimposed pulsating barotropic flow, the maximal exchange regime becomes periodic.

In conclusion, the evidence suggests that it is impossible to create a purely baroclinic stratified exchange flow given the experimental facility described in section 3.1: the initial disequilibrium always results in a propagating pulsating gravity wave, which causes oscillations in the flow rate and influences the interface wave activity.

The generated barotropic oscillations are not only an artefact due to the chosen experimental design: such oscillations appear ubiquitous in environmental and coastal engineering as the basin oscillations between the different portions of a lake (like the lake Lucerne in Switzerland Wuest and Schurter [1988]) or harbor oscillations. In these examples the barotropic component can be caused under the action of surface wind and the periods of oscillations are determined by the geometry of the basins and are of the same order of magnitude of the periods encountered in our experiments. Barotropic oscillations appear also common in oceanography, i.e. in sea straits, or in atmospheric flows as mountain downslope winds, and are caused by astronomical or metereological tides. In these cases, however, the period is not determined by the geometrical features of the basins, as they can be assumed as infinitely large, but by the tide period. These periods are significantly larger than those encountered in our experiments: for example using a water depth of 500 m as typical in the Strait of Gibraltar or in the Faroe Bank Channel, the periods in nature are 50 times larger than the periods in these experiments. Nonetheless, the effects of this barotropic oscillation on the exchange flow and on the interfacial wave characteristics can not be neglected, as also demonstrated by the field observations of Sherwin and Turrell [2005] (see also Hansen and Osterhus [2000]).

Once the origin of the periodic flow rate fluctuations has been explained, the next question which logically arises is what effect has this oscillation in the flow rate on the flow development and on entrainment coefficients. With this aim, results from the experiments 1, 2, 3, 4 and 5 of the PIV measurements are further analyzed and discussed in chapter 5, under conditions of higher Reynolds numbers. In chapter 6 results relative to another serie of experiments are discussed relative to the effect of locally enhanced bottom roughness.

4.5 Summary and conclusions

The flow processes related to stratified exchange flows are generally the result of purely baroclinic effects, and related interfacial instabilities, eventually with superimposed pulsating external forces, which can be provided for example by astronomical or metereological tides or other basin oscillations.

First of all, different starting conditions for the experiment execution have been discussed. The first possibility consists in having the same water level in both the right and left reservoir. In this configuration, there is an unbalanced state prior to the experiment start, as the resultant hydrostatic pressure force on the salt water side is higher than the resultant force on the fresh water side on the separating wall. In the second possibility discussed in section 4.1, the water depth in the fresh water tank is higher, in order to achieve a global balanced equilibrium, before the experiment starts. After the separating wall is removed, the surface water level difference generates a propagating surface wave. Thus, in both cases, additional barotropic forcing is invoked and superimposed onto the baroclinic exchange flow.

Experimental results demonstrate that it is impossible to experimentally generate a purely baroclinic exchange flow.

A simplified model of barotropic oscillations superimposed on exchange flows has been adapted for the presented experimental conditions. The model can predict the flow rate oscillations and the related period, given the geometrical and experimental conditions. The model was finally validated with the experimental results presented here and with the results of Morin and Loewen [2004], revealing good agreement.

5 Experimental results on high Reynolds number exchange flows

In this chapter, results relative to an experimental study of a stratified two-layer exchange flow over a submerged sill with a superimposed pulsating external forcing are presented, under conditions of higher Reynolds numbers, as compared to previous studies of co-current or countercurrent two-layer systems.

The superposed unsteady barotropic component has shown to be responsible for the generation and development of large-scale two-dimensional structures at the interface between the two layers, in form of surges. These large-scale billows have been previously observed also by other authors having similar experimental conditions (Pawlak and Armi [2000], Morin and Loewen [2004], Fouli and Zhu [2005]). The impact of the superimposed pulsating barotropic flow under conditions of higher Reynolds numbers on the development of the shear interface is examined, including results for entrainment coefficients. A fundamental study on hydrodynamic instabilities at the interface of exchange stratified flows is important for modeling the vertical transport of heat, oxygen, nutrients and pollutants in lakes and oceans, as density stratification limits the vertical movement. The chapter is organized as follows. The experimental program is given in section 5.1. A qualitative description of the effect of barotropic oscillations on the baroclinic exchange flow is given in section 5.2. In section 5.3 a quantitative discussion of this effect on interfacial waves is presented, along with an estimation of the length scales (section 5.4). Results relative to entrainment coefficients are finally presented in section 5.5.

5.1 Experimental program

Given the flow scales described in section 3.1 and defining the Reynolds number as $Re = \sqrt{g'H}H/\nu$, there are different possibilities to increase the Reynolds number, namely by increasing the buoyant acceleration g' and/or the water depth H. For this reason, different experiments with different density differences and different water depths have been conducted. It is worth to mention that the change of these parameters will also change the Froude $Fr = \Delta U/\sqrt{g'H}$ and the Richardson $J = g'\delta/(\Delta U^2)$ numbers and of course also the Reynolds number defined with the mean shear velocity ΔU and the shear layer thickness δ as length scale: in table 4.1 the mean values for the different experiments are resumed. From the table, comparing the variation of all these adimensional numbers and the parameters g' and H, it can be noted that only the Reynolds number changes remarcably changing g' or H and a trend can be observed (increasing g' or H, also the Reynolds number is increasing). The Froude number varies very slightly from 0.82 to 1.0, thus remaining in the subcritical range. The Richardson number varies also very slightly from the critical value 0.25 to 0.65. With these arguments, the Reynolds number

Exp.	g' $[cm/s^2]$	H $[cm]$	$Re = \sqrt{g'H}H/\nu$ $[-]$
1	3.09	40.5	45,000
2	3.09	51.0	63,000
3	3.09	59.7	82,000
4	5.81	41.0	60,000
5	5.81	61.0	110,000

Table 5.1: PIV-measurement program with the notations used: g' is the reduced gravity and H is the total water depth. In all experiments a height of the sill crest of $h_c = 20cm$ was used. The sill slope was in all the experiments 0.2 and smooth. Experiments 6 and 7 have been performed using only fresh water. f_{aq} represents the acquisition frequency. Finally, the possibilities illustrated in figure 4.1 for the starting conditions of the experiments are reported.

is chosen as key parameter for the performed experiments. For a more detailed discussion on the flow scales it is referred to section 3.1.2.

In this chapter, experimental results relative to five experiments are presented. The details relative to these experiments are listed in table 5.1. These five experiments are the same as presented in table 4.1 (experiments 1 to 5).

5.2 Visual observations and qualitative description of the flow

When the separating wall is removed, an exchange flow commences. After an initial unsteady period (here of about 1 minute), a maximal exchange regime is established [Armi and Farmer, 1986, Zhu and Lawrence, 1998, Morin and Loewen, 2004] with two internal controls, one at the sill crest, the other at the channel exit. In the last part of the experiment, the flow becomes unsteady again due to the loss of one control (at the channel exit) and the sub-maximal exchange, characterized with a decreasing flow rate, starts. Here, only results of PIV experiments in the steady maximal exchange regime are considered. Figure 5.1 shows a series of instantaneous pictures (with a time interval of 2 s) for experiment 3, in which Rhodamine 6G is mixed in the salt water, three minutes after the beginning of the experiment. The vertical scale is about 60 cm, the horizontal scale 64 cm. At the beginning of this cycle, the interface position at the sill crest increases and is advected downstream (see figure 5.1 (a)). In this region, where the velocity difference between upper and lower layer is increasing due to the accelerating flow down the sill, the formation of a large-scale billow can be observed at approximately $x/L = 0.25$ (see figure 5.1 (b), (c) and (d)). This billow is advected downstream and grows further, entraining a lot of fresh water into the core of the billow, at the same time carrying salty water from the lower layer to higher water depths, where it can be mixed with the fresh water in its core (see figure 5.1 (e) and (f)). Until this point, a very sharp interface is created on the front (on the right) of the billow, where small scale KH billows are generated (cf. figure 5.2 (a)). In figure 5.1 (g) the billow starts to collapse creating a more sharp interface upstream, characterized again by the formation of small scale KH billows. Meanwhile the interface at the sill crest increases again and produces at the front

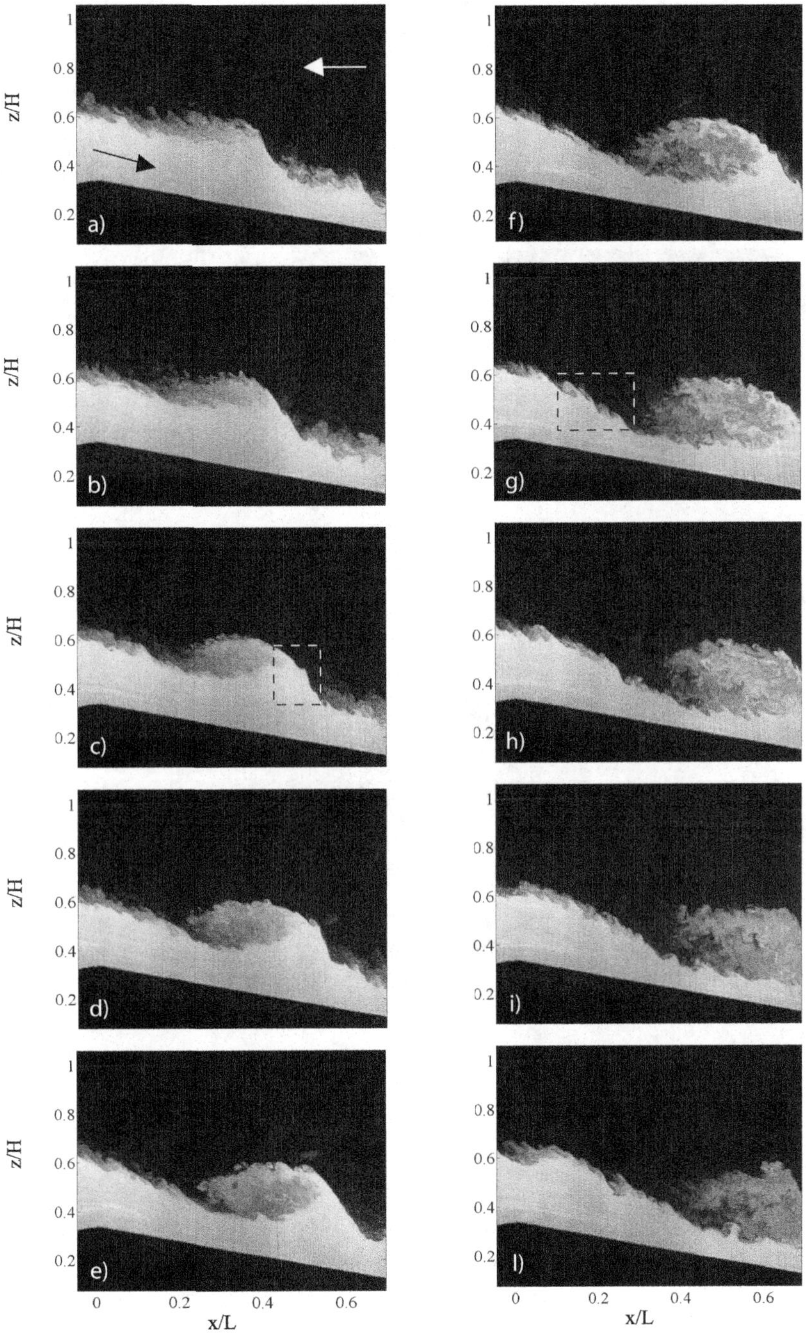

Figure 5.1: Series of images taken 3 minutes after the beginning of the experiment (experiment 3) with Rhodamine diluted in the salt water layer flowing from the left to the right. The interval between successive images is 2 s. The image is 56 cm vertical and 70 cm horizontal. In the pictures (c) and (g) the dashed squares represent the zoomed zones in figure 5.2 (a) and (b), respectively.

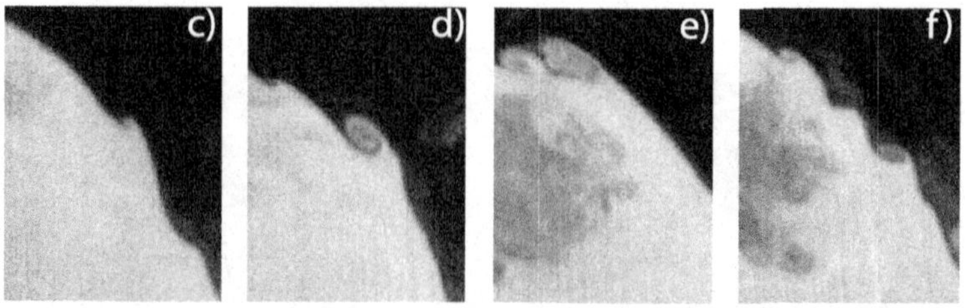

(a)

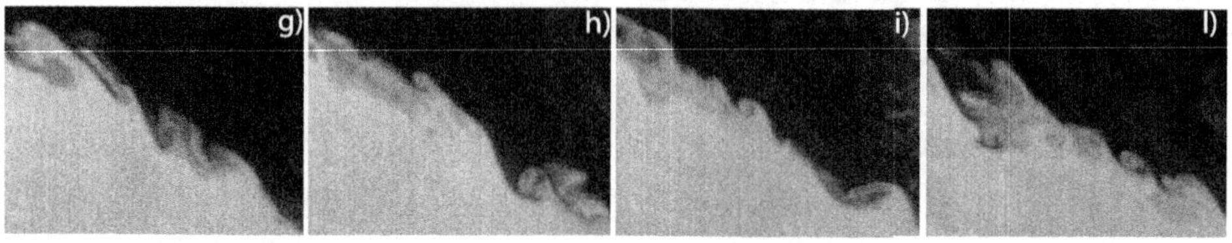

(b)

Figure 5.2: (a) Zoom in of the front of the billow in figure 5.1 (c), (d), (e) and (f), and (b) of the sharp interface in the wake direct behind the billow in figure 5.1 (g), (h), (i) and (l). The image is (a) 10 *cm* (vertical) by 6 *cm* (horizontal) and (b) 6 *cm* by 15 *cm*, respectively.

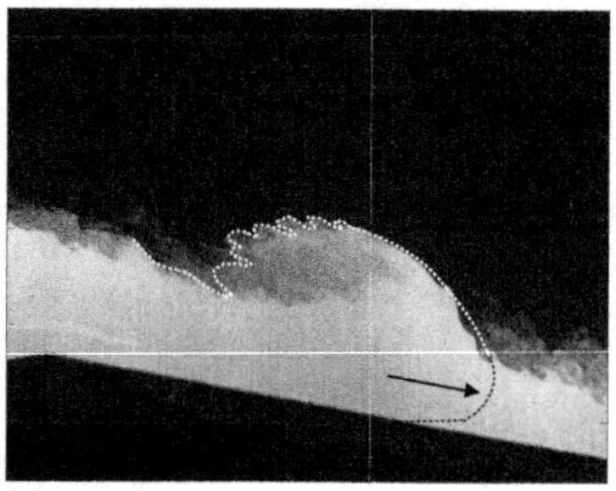

(a)

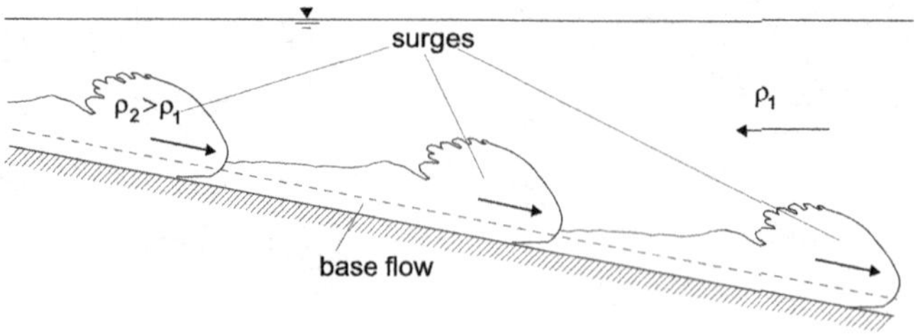

(b)

Figure 5.3: Physical explanation of the large-scale wave mechanism observed during the experiments. (a) Instantaneous image (picture (c) in figure 5.1) with a surge-like flow sketched onto the picture (dashed lines). (b) Schematization of the exchange flow down the sill with periodic surge formation to balance the barotropic generated overflows.

of the wave further KH instabilities (see figure 5.2 (b)), while the old billow continues to decay as it is stretched and further advected downstream. The high interface front at the sill crest is then advected down the sill (figure 5.1 (l)) and the phenomena starts again from the beginning as in figure 5.1 (a). The KH instabilities generated at the sill crest are advected in both down and upstream directions.

In summary, the flow region investigated in this study can be divided into three zones: the first region (region 1) on the left of the sill crest (at $-0.15 < x/L < 0$, cf. figure 5.1), characterized with a thin shear and density layer, in which small scale instabilities can be generated, but are advected from the right to the left. The second region (region 2, at $0 < x/L < 0.2$, cf. figure 5.1)) is characterized with a developing and growing shear and density layer, in which typically KH instabilities are generated, grow and eventually collapse within this region. They are normally advected downstream from the left to the right. In the third region (region 3) the large scale billows are generated with a low frequency (at $0.2 < x/L < 0.55$, cf. figure 5.1).

The large scale billows, which are periodically generated during the maximal exchange regime have been seen in such laboratory experiments also by previous authors (e.g. Morin and Loewen [2004], Fouli and Zhu [2005]), although their generating mechanism has apparently not been identified and analyzed in detail. From experimental observations, with dye diluted in the lower layer, the life period of these large scale interfacial instabilities is of the same order of the period of the barotropic flow rate oscillations, depending on the Reynolds number. A possible physical explanation for the observed large scale billows consists in considering them as a series of surges that lead to a pulsating density current in the lower layer, superposed to a base flow (see figure 5.3). These surge-like elements are generated periodically at the sill crest to assure that the barotropic induced overflow can be exchanged from one reservoir to the other. These surges act very similar to thermals on a sloping bottom (see for example Turner [1968]). Also, there is a strong similarity of these pulsating surge-like flows with roll waves in single-layer flows, which are explained in Baines [1989] as a sequence of periodic hydraulic jumps propagating down a slope and which have shown to be spatially periodic stable solutions of the linear stability analysis of a uniform, steady flow with bottom slope and roughness (Whitham [1967]). These type of waves have been observed in atmospheric flows, for example in downslope windstorms in Boulder, Colorado (Scinocca and Peltier [1989]).

The next section will adduce as evidence the physical explanation of the large scale billows which has been given above.

5.3 Interfacial wave characteristics

The time series of the interface position were used to generate a wave characteristic plot (figure 5.4), from which the wave activities can be extrapolated. Two sets of waves propagating in opposite direction are generated at the sill crest: the steeper slope of the wave characteristic line of the waves traveling downstream (continuous lines in figure 5.4, Exp. 4) as compared to the slope of the characteristic lines of the waves propagating upstream (dashed lines in figure 5.4, Exp. 4), suggest that the waves propagating upstream are smaller in amplitude and travel slower compared to the waves propagating downstream. The large breaking waves observed during the maximal exchange are shown from the

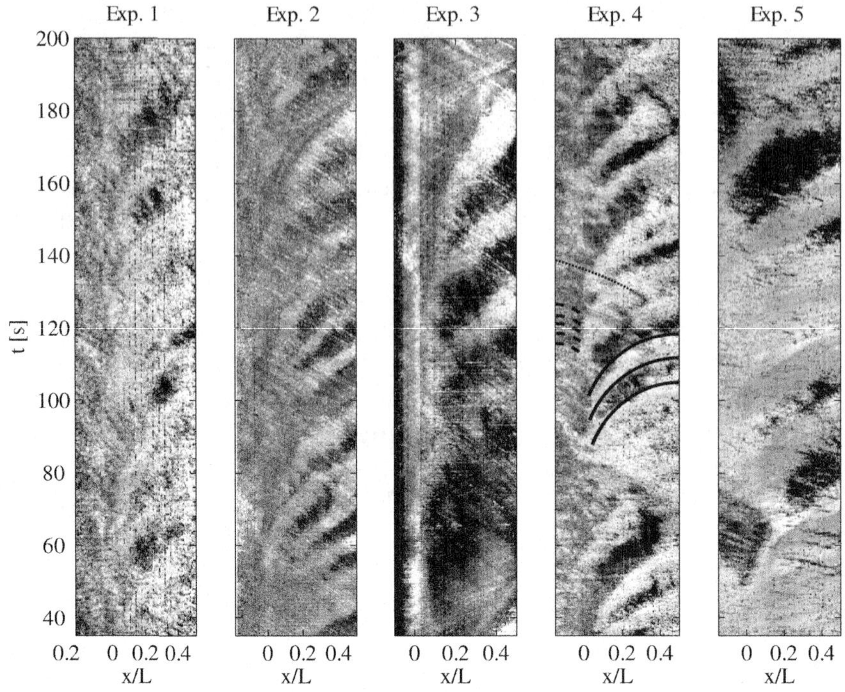

Figure 5.4: Wave characteristic plot of experiments 1, 2, 3, 4, 5. Dark zones are locations at which the instantaneous interface is higher as the mean interface, bright zones regions where the instantaneous interface is smaller as the mean value.

plot to be generated with a period of approximately 20 s for all the experiments (cf. the gap between the continuous lines in figure 5.4, Exp. 4), within $0 < x/L < 0.15$. Beyond these larger front, the generation of smaller KH at their steep fronts can be noted (see dotted line in figure 5.4, Exp. 4). Spectral distributions of the velocity fluctuations were calculated from the velocity fields obtained from the PIV measurements to have an overview on the scales dominating the flow. These energy spectra gives only qualitative estimations of spectral slopes and behaviors due to the relatively low acquisition frequency and spatial resolution of the PIV experiments. Figure 5.5 shows representative energy spectra of the velocity fluctuations of the main velocity component (continuous line) and of the vertical velocity component (dashed line) at $x/L = 0.1$. Slopes of $-7/3$ and -2 are shown for comparison along with the typical turbulent spectral slope of $-5/3$. The spectral slope is a measure of the rate of energy transfer between the scales since the Reynolds number is large enough. For all experiments the first part of the spectrum is better described by a $-7/3$ slope, while the central part is better described by a -2 slope. The classical $-5/3$ slope characterize more the last part of the spectrum until frequencies of $2 \cdot 10^0$. An important issue in the study of stratified flows is to separate the large scale generated frequencies caused for example by internal waves, and the contribution by the buoyancy driven small scales. Low frequency ranges in stratified flows have shown to be better described by -2 and $-7/3$-slopes, for frequencies usually lower than the buoyancy frequency N [Lien and Dairiki, 1998]. In this zone of the spectrum,

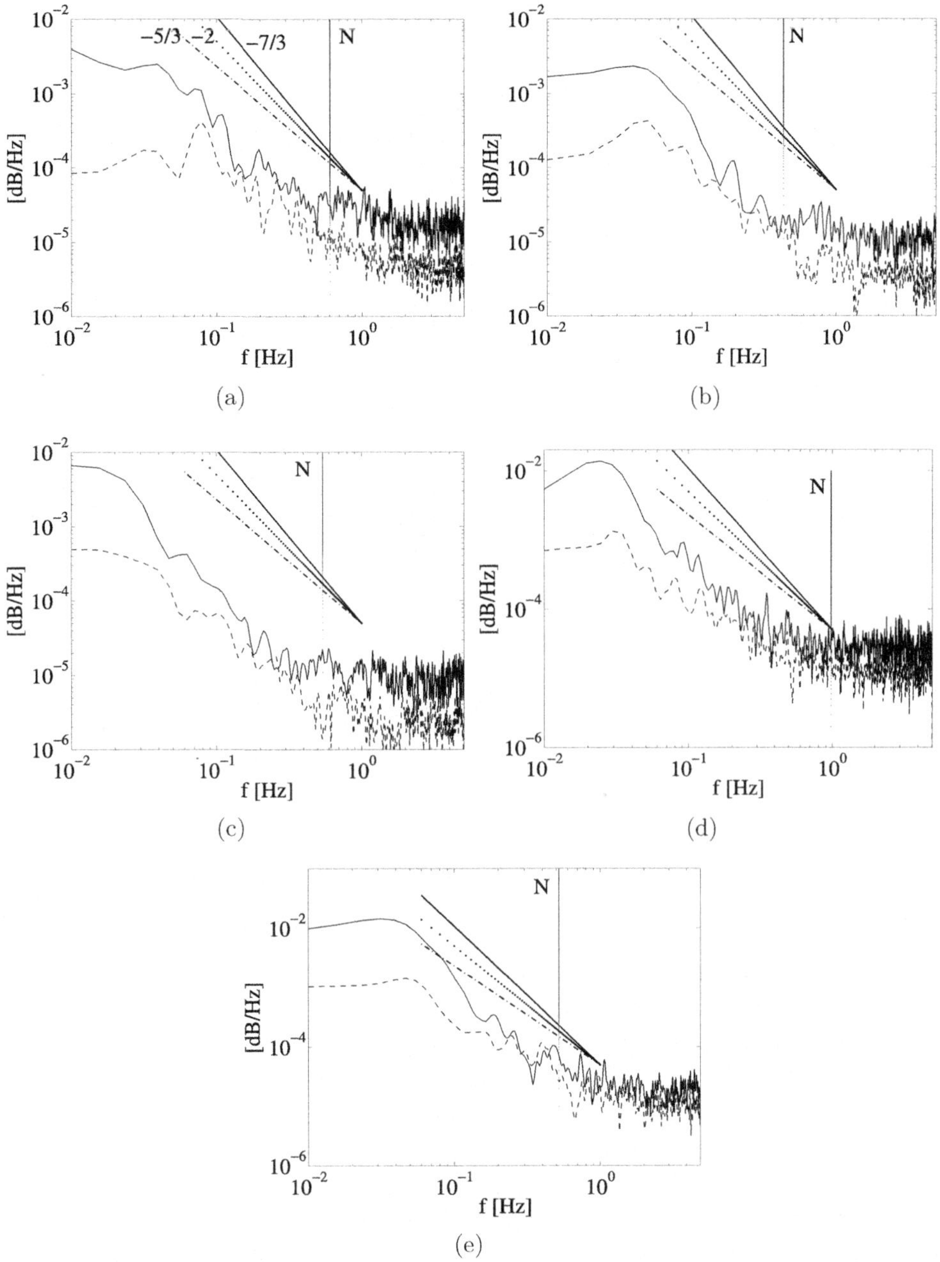

Figure 5.5: Spectral distribution of the velocity fluctuations in streamwise (continuous line) and in vertical direction (dashed line) at the interface position at $x/L = 0.2$ for Experiments 1 (a), 2 (b), 3 (c), 4 (d) and 5 (e). Spectral slopes of $-7/3$, -2 and $-5/3$ are plotted for comparison along with the (bulk) buoyancy frequency $N = g'/\delta$. The spectra of the vertical velocity fluctuations are one order of magnitude lower than those of the horizontal component for low frequencies. Moreover, the two spectra coalesce earlier for the experiments with higher buoyant acceleration ((d) and (e)).

the velocities are anisotropic, as also demonstrated by the difference in the spectrum of the horizontal and vertical components of the velocity: first, the slopes in the spectra of the vertical component of the velocity (dashed lines in figure 5.5) for all the experiments is from the beginning better described by a $-5/3$ slope. Second, the effect of buoyancy can be seen in the vertical velocity fluctuations at the larger scales where the energy is approximately one order of magnitude less than that of the streamwise fluctuations. For increasing frequencies, the velocities become nearly isotropic and the spectral slopes for the horizontal component of the velocity decrease to the $-5/3$-slope, characterizing the small scale buoyancy dominated turbulence. The data relative to the presented spectral slopes will be discussed also in section 5.5. Despite the fact that at these scales the limits of the PIV averaging becomes significant, the energy in the last part of the spectrum shows a distinct separation, which is very remarkable for experiments 1, 2 and 3 and is much more weak for experiments 4 and 5 characterized with a higher buoyancy, where the spectra of the streamwise and vertical velocity components coalesce very early, at approximately $f \sim N$. For the first three experiments the gap between the two spectra remains nearly constant for all the range of frequencies. This is probably due to the increasing Reynolds number (see table 4.1), which inhibits the stabilizing effect of buoyancy (see table 4.1). For comparison, the (bulk) buoyancy frequency $N = g'/\delta$ is plotted in figure 5.5.

The streamwise rms velocity fluctuation u' are plotted versus streamwise direction in figure 5.6 (a), normalized with the local velocity difference ΔU. This ratio gives the response of the velocity fluctuations to the velocity difference increase due to the acceleration down the sill. In experiment 1 the ratio increases, while in experiment 4 the ratio remain constant on a value of ≈ 0.28. This different tendency between experiment 1 and 4 can be explained by the higher restoring force given by the higher buoyancy in experiment 4, which works for stabilization. The effect of the higher water depth in experiments 2, 3 and 5 is to reduce the increasing rate of the gap between the two velocity scales as their development in streamwise direction remains nearly constant. Figure 5.6 (a) shows also that an increase in the water depth and in buoyancy, i.e. the more the Reynolds number increases, the more the response of the velocity fluctuations to the acceleration down the sill decreases. Increasing the Reynolds number the mean values of the ratio changes from 0.4 to 0.2. Koop and Browand [1976] and Pawlak and Armi [2000] have reported a decay of this ratio in the longitudinal direction of $x^{-3/4}$ as an indicator of collapse of the initial turbulence in the shear layer. This is not observed here, as the increasing Reynolds number in the accelerating, rapid growth region right to the sill crest results in an input of energy, thus keeping the ratio nearly constant.

The mean shear layer thickness was defined as $\delta = h|_{\overline{u_1}=0.85} - h|_{\overline{u_2}=0.85}$, where $h|_{\overline{u_i}=0.85}$ is the water depth at which the velocity in the upper ($i = 1$) or lower ($i = 2$) becomes 85% of the maximal velocity in that layer. The mean shear layer thickness along the longitudinal direction is plotted in figure 5.6 (b). The increase for experiment 1 and 4 is linear and almost equal for both. It is also worth noting that an increase of the water depth causes the shear layer thickness to grow faster, while an increase of the buoyant acceleration results in a slight slower growth of the shear layer. This is obviously due to the buoyancy, inhibiting the vertical motion. The bulk Richardson number is here defined using the velocity shear layer thickness, $J = g'\delta/\Delta U^2$, (see figure 5.6 (c)). For all the experiments, it remains nearly constant and it ranges from 0.2 for experiments 1 and

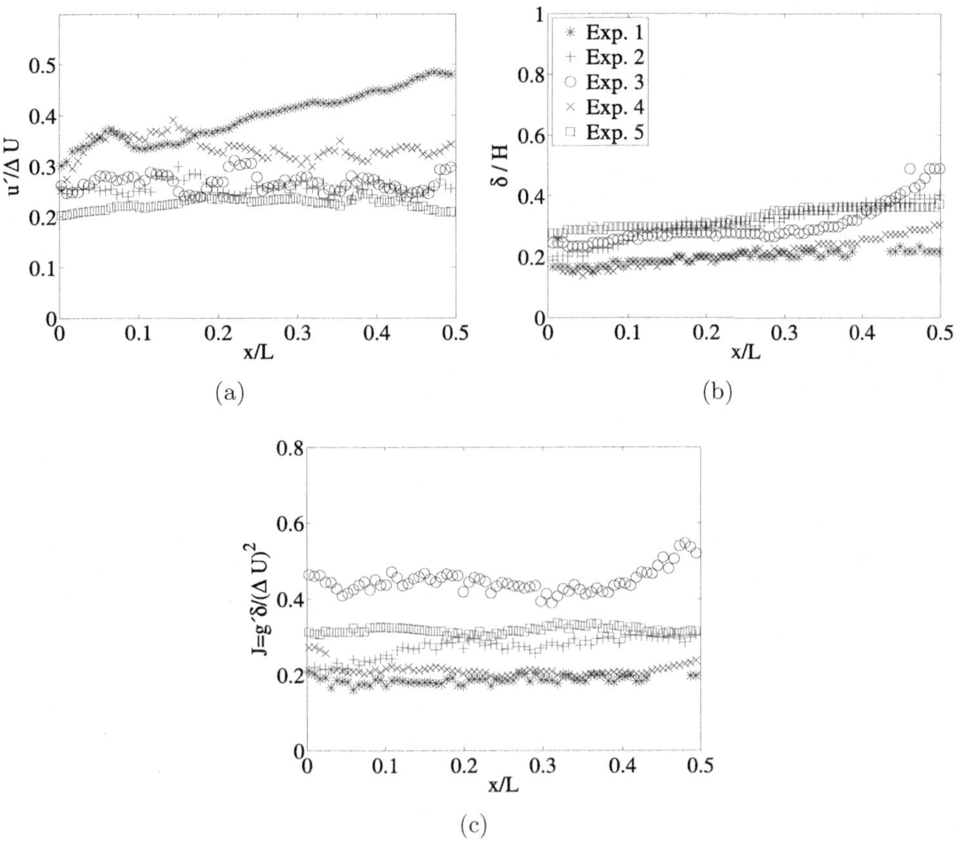

Figure 5.6: Time averaged quantities for all the performed experiments. (a) Maximal r.m.s. of the streamwise velocity component normalized with the local maximal velocity difference vs. x. In contrast with the results of Koop and Browand (1976) and Pawlak and Armi (2000) the ratio remains constant along the longitudinal direction and does not show the decay observed previously of $x^{-3/4}$. This suggests that all the region treated in this study is a turbulence growth region. (b) Shear layer thickness and correspondent slopes of lines representing the fitted data and (c) bulk Richardson number.

4 to 0.5 for experiment 3. These averaged quantities of the mean shear layer thickness and Richardson number are interesting for the discussion on the length scale estimation of the large scale interfacial structures.

The streamwise rms velocity fluctuation u'_{rms} are plotted versus streamwise direction in figure 5.6 (a), normalized with the local velocity difference ΔU. This ratio gives the response of the velocity fluctuations to the velocity difference increase due to the acceleration down the sill. In experiments 1 and 4 there is an abrupt increase of this ratio at the sill crest, then, in experiment 1, it continues to increase, while in experiment 4 the ratio remain constant from $x/L \sim 0.1$ on a value of 0.33. This different tendency between experiment 1 and 4 can be explained by the higher restoring force given by the higher buoyancy in experiment 4, which works for stabilization. The effect of the higher water depth in experiments 2, 3 and 5 is to reduce the increasing rate of the gap between the two velocity scales as their development in streamwise direction show a moderate but an overall increase. Figure 5.6 (a) shows also that an increase in the water depth and in buoyancy, i.e. the more the Reynolds number increases, the more the response of the velocity fluctuations to the acceleration down the sill decreases. Increasing the Reynolds number the mean values of the ratio changes from 0.2 to 0.4. This has repercussions also on entrainment quantities, which are presented in detail in section 5.5. Koop and Browand [1976] and Pawlak and Armi [2000] have reported a decay of this ratio in the longitudinal direction of $x^{-3/4}$ as an indicator of collapse of the initial turbulence in the shear layer. This is not observed here, as the increasing Reynolds number in the accelerating, rapid growth region right to the sill crest results in an input of energy, thus keeping the ratio nearly constant.

The mean shear layer thickness along the longitudinal direction is plotted in figure 5.6 (b). The shear layer thickness undergoes an abrupt increase near the sill crest for the experiments with a water depth $H = 40\ cm$ at $x/L = 0.1$ for experiment 1 and at $x/L = 0$ for the experiment with the same water depth but with a higher buoyant acceleration (experiment 4). The increase for these two experiments is then linear and almost equal for both. Experiments 2, 3 and 5 do not present an abrupt increase of the shear layer and the curve is linear, probably because the impact encountering the sill crest is not so high. The shape of the line fitting the data for the shear layer thickness increases for the experiments 1, 2, 3, increasing the water depth and taking the buoyant acceleration constant, but it is decreasing increasing the buoyant acceleration and taking the same water depth. The bulk Richardson number, here defined using the velocity shear layer thickness instread of the density layer thickness, $J = g'\delta/\Delta U^2$, (see figure 5.6 (c)) shows an abrupt increase in longitudinal direction at the same locations as the abrupt increase in the shear layer thickness. For experiments 1 and 4 it increases from $x/L = 0.3$. In Experiments 2, 3 and 5, it remains constant. These averaged quantities of the mean shear layer thickness and Richardson number are interesting for the discussion on the length scale estimation of the large scale interfacial structures.

5.4 Estimation of length scales

For the estimation of the length scales of these structures, the temporal correlation functions $R(\tau)$ of the horizontal (see figure 5.7 (a)) and vertical (see figure 5.7 (b)) velocity components has been determined. The length scales were then calculated as product between the time scale $T_I = \int_0^{\tau_I} R(\tau)d\tau$, with τ_I being the first reached minimum value of the correlation function, and a convective velocity scale, here assumed to be half of the velocity difference $\Delta U/2$.

One fundamental issue in the investigation of interfacial instabilities in stratified exchange flows is the determination of the key parameters governing the generation, the growth and development of these instabilities. They are the result of an interplay between buoyancy and the mean shear velocity for a constant slope of the sill. As the flow treated here is also governed by barotropic forces, we expect the water depth is playing a crucial role in determining the size of these instabilities. Thus, we argue that the key parameter governing the generation and development of these surge-like waves is the water depth H. The buoyancy seems to have a secondary importance in influencing the size of these instabilities at the interface. In this sense, a Reynolds number defined using the velocity

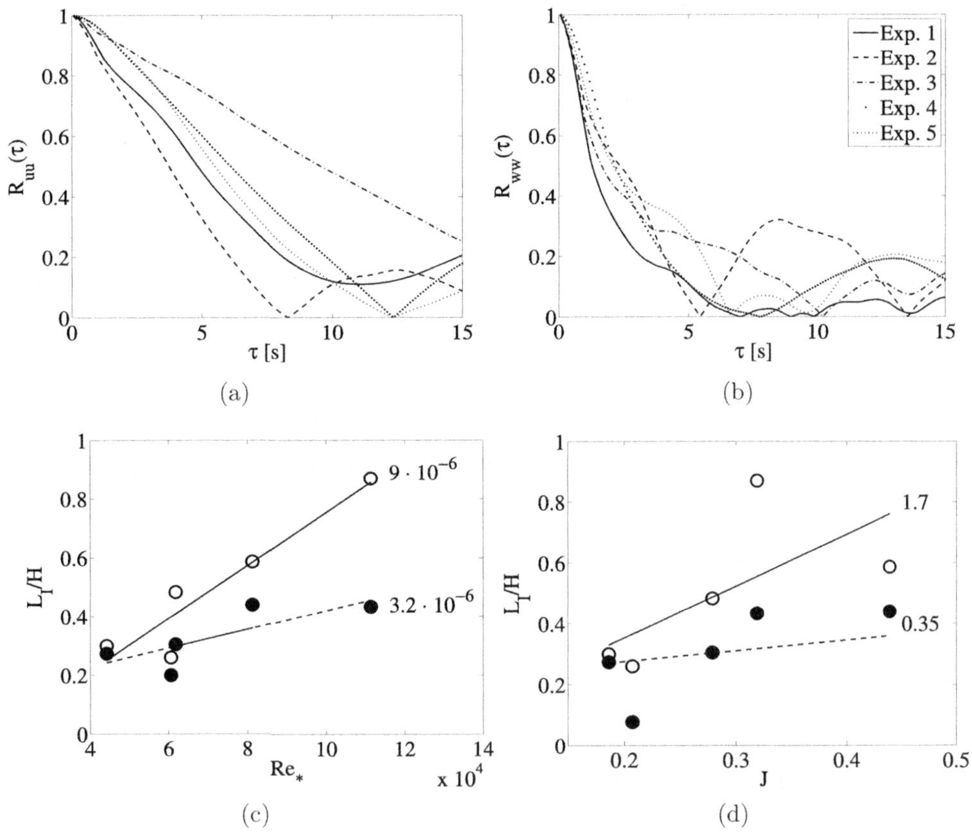

Figure 5.7: Estimation of the length scales of the large structures observed in the experiments. Temporal correlation functions of the horizontal (a) and for the vertical (b) velocity components for all the experiments. (c) Estimated length scales from the correlation functions of the horizontal (empty symbols) and of the vertical (filled symbols) length scales vs. the overall Reynolds number $Re_* = \sqrt{g'H}H/\nu$ and (d) vs. the overall bulk Richardson number $J = g'\delta/(\Delta U)^2$. The integral length scales were obtained integrating the correlation function until it reached its first minimal value.

scale $\sqrt{g'H}$ and as length scale the water depth H, i.e. $Re_* = \sqrt{g'H}H/\nu$, should be an appropriate parameter to illustrate the development of the length scales of such waves in a stratified exchange flow with superposed pulsating barotropic forcing, as firstly the effects of changes in the water depth are taken into account and, secondly, the effects of buoyancy.

In figure 5.7 (a) the horizontal (empty symbols) and vertical (filled symbols) length scale estimations are plotted versus the Reynolds number. Both the vertical and horizontal estimated length scales show overall increasing values with increasing Re_*. If the data are fitted with a linear curve, a slower increase for the vertical scales can be noted. This is obviously due to the buoyancy, which inhibits vertical motion. Interesting to note is that increasing the buoyant acceleration (experiments 1 to 4 and experiments 3 to 5) the estimated length scales are not really increasing but actually remain constant or decrease. This is observed for values of the Reynolds number smaller then approximately

$0.6 \cdot 10^5$. At this value, there is an abrupt increase of the size of the scales with increasing Reynolds number, and the buoyancy seems to alleviate his stabilizing effect. We thus propose a relation between the Reynolds number Re_* and the length scales of the large scale surges as follows

$$\frac{L_I}{H} = \kappa Re_*$$

were κ is $0.33 \cdot 10^{-5}$ for the vertical scales and $0.9 \cdot 10^{-5}$ for the horizontal scales. More precise estimations of the value of the constant κ should be achieved through additional experiments. For all experiments, the horizontal scale is larger than the vertical scale. The vertical scales show to be approximately 1/3 of the total water depth for experiments 1 and 2, while they are roughly 1/2 for experiment 3 and 5 and 1/4 for experiment 4. Also the results for the horizontal length scales show consistence with the experimental observations presented in section 5.2.

If the scales are plotted vs the Richardson number no firm conclusions about a tendency are possible, as shown by the larger difference between the experimental data and the linear fit in figure 5.7 (b) as compared to figure 5.7 (a). The bulk Richardson number is a measure of the stabilizing effect of buoyancy relative to the destabilizing effect of the mean shear. Changes in the bulk Richardson number are primarily the result of changes in the buoyant acceleration. As mentioned above, a change in the buoyant acceleration produces both changes in the velocity difference and in the shear layer thickness. All these parameters are included in the definition of the bulk Richardson number and relative changes between these scales are complicated and difficult to predict so that this number is much more sensitive to changes in both the density difference and the water depth, as also demonstrated in figure 6.5 (c). The generation of the large-scale surges in the lower layer, on the other hand, is triggered by the pulsating barotropic component and is influenced by changes in the water depth, rather than by the buoyant acceleration. This explains why the Reynolds number is more appropriate for predicting the growth of the length scales compared to the Richardson number.

5.5 Mixing and entrainment

Given the velocity field data, estimates of the entrainment coefficients can be obtained. Following Ellison and Turner [1959], the entrainment coefficient is defined as

$$\alpha_e = \frac{1}{V}\frac{dq}{dx} \quad \text{with} \quad V = \frac{\int u^2 dz}{\int u dz} \tag{5.1}$$

where q is the flow rate per unit width in the layer of interest and V is a representative velocity for the layer. Figure 5.8 (a) shows the entrainment coefficients for all the experiments and the higher values are found in the region $0 < x/L < 0.2$. The overall lowest entrainment coefficients are observed for the experiments 1 and 4, which are of the same order of magnitude as the entrainment coefficients found by Morin and Loewen [2004] and Fouli and Zhu [2005], with the highest values approaching the value 0.1. The entrainment coefficients for the experiments 2, 3 and 5 pass this level and reach maxima of 0.13, 0.15 and 0.25, respectively which are very high values compared to that observed previously in stratified exchange and arrested flows (see the data of Pawlak and Armi [2000] in figure

5.8 (a)) and in density currents. In figure 5.8 (b), entrainment coefficients α_e of the here performed experiments along with the results obtained by Ellison and Turner [1959] are plotted versus the bulk Richardson number $Ri = g'\delta \cos\alpha/(\Delta U)^2$, as it was defined by Ellison and Turner [1959]. With exception of experiment 5, all the other experiments have values of the entrainment coefficients comparable with the results obtained by Ellison and Turner [1959]. The observed increased entrainment coefficients for increasing Reynolds number is surprising, as generally higher Reynolds numbers leave to reduced entrainment coefficients. This is obviously not the case here, because the first contribution to the increase of the Reynolds number is given by an increased water depth. The higher water depth is responsible for a higher barotropic component and thus is related to larger sizes of the surge-like waves (cf. figure 5.7). These large scale structures deliver the primary contribution to the entrainment coefficients. This also show consistence with figure 5.8 (c), which will be discussed in the following.

To analyze the link between the buoyancy dominated large scales, which represents the primary contribution to u', and the smaller-scale fluctuations, associated with entraining motions, the ratio of entrainment velocity $u_e = \alpha_e V_q = dq/dx$ to u' plotted in x direction represents an effective comparison between these scales [Pawlak and Armi, 2000]. The entrainment velocity is representative of the entrainment motions. For all the experiments there is a rapid increase in the initial region (see figure 5.8 (c)), where u_e is a significant fraction of u' (up to $x/L \sim 0.1$). Further downstream, the ratio drops significantly as the gap between the scales increases. These results are also consistent with the evolution of the Richardson number J. Pawlak and Armi [2000] observed for this ratio an increase in the initial, rapid growth, developing region and a drop farther downstream as the gap between the scales increases. This is not the case here, where obviously the gap is decreasing moving downstream from the sill crest. This suggests that the region studied here is overall a growth region.

The variation of the ratio between the entrainment velocity u_e and the rms of the velocity fluctuations as a function of the Richardson number based on the large scale of turbulence was investigated by numerous authors. For example, Ellison and Turner [1959] observed a dependence of $u_e/u' = Ri^{-1}$ for large Richardson numbers and of $u_e/u' = Ri^{-3/2}$ for moderate Richardson numbers (see also Baines [1975], Fernando [1991], Crapper and Linden [1974]), where the Richardson number was defined using the reduced gravitational acceleration g', the length scale associated with the wave number k from the energy spectra and the rms of the velocity fluctuations in streamwise direction u'. Cotel and Breidenthal [1997] have suggested a model based on vortex persistence to account for the varied dependence on the Richardson number, the Schmidt number and the Reynolds number. Pawlak and Armi [2000] suggested a relation of $u_e/u' = Ri^{\frac{n+1}{2n+6}}$, where n is the spectral slope. This relation gives the dependence of the entrainment velocity on the bulk properties of the turbulence given a spectral shape along with the limiting effect of buoyancy on the entrainment. The relation by Pawlak and Armi [2000] matches well the prediction of Ri^{-1} by Cotel and Breidenthal [1997] using the observed spectral shape of $-7/3$ characterizing the low frequency region of the spectra presented in figure 5.5. In figure 5.8 (e) this ratio is plotted versus the Richardson number $Ri = g'l'/u'$ for all the experiments: for g' the constant mean values for each experiment were used (see table 4.1) while l' was calculated using the convective velocities of $\Delta U/2$ and the

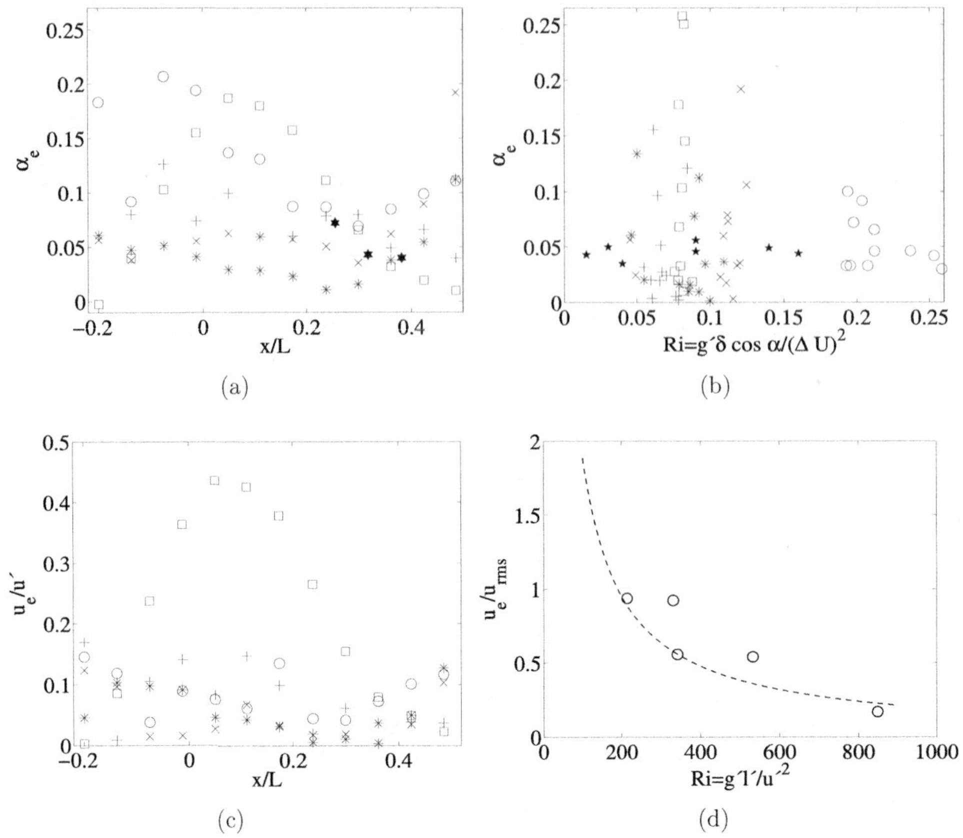

Figure 5.8: Downslope entrainment coefficient of lower-layer fluid for exp. 1 ($*$), 2 ($+$), 3 (o), 4 ($\times$), 5 ($\Box$) compared with the results of Pawlak and Armi (2000) ($\star$) (a), and entrainment coefficients compared to the coefficients obtained by Ellison and Turner (1972) ($\star$) (b). Entrainment velocity normalized by the maximum rms velocity fluctuations in streamwise direction vs. x (c). Ratio between the entrainment velocity u_e and the rms of the velocity fluctuations as a function of the Richardson number based on the large scales of turbulence for all the performed experiments (d), showing a good agreement with the vortex persistence model of Cotel and Breidenthal ($u_e/u' \propto Ri^{-1}$). The dashed line represents Ri^{-1}, symbols represent the performed experiments.

dominant frequencies observed in the spectra of the horizontal velocity fluctuations for each experiment. If the data are fitted with an exponential function, an exponent of -0.1 for the Richardson number can be obtained, which compares well with the vortex persistence model of Cotel and Breidenthal [1997].

5.6 Summary and conclusions

In this chapter, results of physical experiments have been presented, in order to understand the effects related to higher Reynolds numbers.

From dye visualizations in the experiments it could be observed that not only KH instabilities at the interface between the two layers were generated (as typically and just largely observed in previous studies), but also a new type of instability in form of large scale billows. These billows take the form of pulsating surges to balance the overflow from one water body to the other. This type of instability at the interface has been observed in prior studies [Morin and Loewen, 2004, Fouli and Zhu, 2005], but has not been analyzed in detail. The period of generation of these instabilities has shown to be in the same order of magnitude of the period of oscillation of the net flow rate.

Results for five experiments have been examined more in detail to show the impact of the flow pulsation on the development of the flow and on entrainment coefficients. From the time analysis the period of the flow rate fluctuations could be recognized again in the velocity fluctuation spectra. Slopes of $-7/3$ have been reported for smaller frequencies, along with slopes of $-5/3$ in the last part of the spectrum for higher frequencies. These different slopes suggest the presence of two-dimensional structures which decay in higher frequencies as the turbulence becomes more isotropic. The spectral distributions of the velocity fluctuations in horizontal and vertical direction also show differences of one order of magnitude in the energy values, respectively. This is more evident for the experiments with a lower Reynolds number (cf. figure 5.5).

The results of averaged velocity fields reveal an increasing rate of the velocity shear layer thickness with increasing Richardson number J (cf. figure 5.6) and that the overall region studied is characterized with growing turbulence as the Richardson number increases in downstream direction in experiments 1, 2 and 4, while in experiments 3 and 5 the Richardson number remains constant in longitudinal direction. The ratio $u'/\Delta U$, which gives the response of the velocity fluctuations to the velocity difference increase due to the acceleration down the sill, has revealed an overall increase in longitudinal direction, which is different from the results by Koop and Browand [1976] and Pawlak and Armi [2000], who reported a decaying rate in longitudinal direction of $x^{-3/4}$.

The vertical and horizontal length scales of the observed large scale breaking waves has been estimated using the time scale obtained from the correlation functions of the velocity fluctuations and the convective velocity $\Delta U/2$ (cf. figure 5.7). Both the vertical and horizontal scales are increasing with increasing water depth, but slightly decreasing with increasing buoyant acceleration for $Re_* = \sqrt{g'H}H/\nu < 60,000$. This Reynolds number has shown to be the key parameter governing the generation, the development and growth of the interfacial instabilities for stratified exchange flows with superposed pulsating barotropic component. Considering the overall behavior, it could be concluded that for increasing Re_*, the scales show an increase and the relation $L_I/H = \kappa Re_*$ has been proposed, κ being 0.9 for the horizontal scales and 0.33 for the vertical scales. A similar behavior has been also observed for the values of the length scales plotted versus the Richardson number, where a very similar relation between the scales and the Richardson number is found as compared to the relation obtained using the Reynolds number.

Entrainment coefficients have shown to be similar to the values of Ellison and Turner [1959] but larger than those reported by Pawlak and Armi [2000], Morin and Loewen [2004] and Fouli and Zhu [2005]. The results show higher entrainment coefficients for increasing Reynolds number, which is surprising as generally the opposite behavior has been observed. This is due to the fact that the increase of the Reynolds number is caused by higher water depths, which increases the barotropic oscillations and so the size of the surge-like waves. The ratio between entrainment velocity and maximal rms values of the streamwise rms velocity fluctuation plotted versus the Richardson number $Ri = g'l'/u'^2$, as defined in Pawlak and Armi [2000] show a decrease with increasing Richardson number with a dependence of Ri^{-1}, which is consistent with the predictions of Ellison and Turner [1959] and with the vortex persistence model of Cotel and Breidenthal [1997].

6 Experimental results on locally enhanced bottom roughness

In natural water bodies, in the oceans and in the atmosphere, bottom roughness is ubiquitous and is a key parameter in influencing the flow behavior and the related entrainment and mixing processes. Indeed, most of the previous studies dealing with boundary layer dominated flows focused their attention on the subsequent effects on the hydraulics of these flows (Zhu and Lawrence [1998], Zaremba and Pieters [2003], Gu and Lawrence [2005]) and the related entrainment and mixing (Dallimore and Ishikawa [2001], Fernandez and Imberger [2007]) rather than analyzing their effect on the interfacial wave activity.

This chapter is aimed at investigating the effect of bottom friction on the generation and growth mechanism of the large-scale, surge-like structures at the interface between the two layers. Such surge-like waves are observed in oceanography (Sherwin and Turrell [2005] see also Hansen and Osterhus [2000]) and in atmospheric flows, for example in downslope windstorms in Boulder, Colorado (Scinocca and Peltier [1989]).

Thorpe [1983] performed experiments of linearly stratified flows down a slope over a rough floor, where the roughness elements were represented by square bars set at regular intervals, studying both the configurations of k-type and d-type as defined in Perry and Joubert [1969], the first type characterized by dense roughness elements with a high interaction with the overlaying layer, and the second type characterized by a lower interaction with the overlaying layer, as the roughness elements act more as single disturbance bodies. In his experimental study, Thorpe [1983] observed the generation of radiating internal gravity waves at the initial stage of the flow and reported turbulent mixing behind the bars due to flow separation but low interaction with the interface between the two layers. He showed that this turbulent layer spread vertically less rapidly than the internal waves and that the rate of spread depends on the separation of the bars: the nearer the bars, the higher was the interaction between the bottom generated and the interface generated turbulence. This interaction was additionally enhanced by using an array of square cubes instead of bars.

In this chapter, a specific goal is the consideration of the combined effects of bottom generated turbulence and internal baroclinic generation of vorticity. The experimental results presented here will show that bottom friction has a significant impact on the developing mechanism of these interfacial instabilities showing a damping effect on their growing and collapsing mechanisms and that their generation frequency depends linearly upon their size. Moreover, dense bottom friction has shown to reduce entrainment coefficients if a sparser grid (the d-type) of roughness elements is employed.

The chapter is organized as follows. In section 6.1 a discussion on the different sources of turbulence production is presented and three different parameters describing the ratio between these sources are defined. An estimation of the bottom roughness coefficients along with a comparison with hydraulic theory results is given in section 6.4. The effect of the bottom roughness on interfacial waves is discussed in section 6.6, while effects on the turbulence characteristics and on the entrainment coefficients are given in sections 6.7 and 6.8 respectively.

6.1 Vorticity production in a two-layer stratified flow over a rough sill

The definition diagram for the vorticity generation in a gradually evolving two-layer flow over downslope is given in figure 6.1. In such a flow, different turbulence mechanisms are responsible for vorticity production: a first source of vorticity is given by the vortex stretching due to the variation in height of the channel because of the sill placed on the bottom. A second source is represented by the baroclinic generation of vorticity, given by the inclination of the interface. In addition, the enhanced bottom roughness is contributing to the vortex stretching as an additional source of turbulence production. It would be useful to relate the turbulence produced at the bottom with the turbulence produced by the density and velocity gradients and thus, we define some parameters which can express the ratio between the magnitude of each contribution. The starting point for this analysis is the vorticity equation for a two-layer stratified flow which reads (Baines [1975])

$$\frac{D\zeta}{Dt} = \zeta \cdot \nabla \mathbf{u} + \frac{\nabla\rho \times \nabla p}{\rho^2} + \nu\triangle\zeta + F.S. \tag{6.1}$$

where D/Dt denotes the Lagrangian derivative, $\mathbf{u} = (u, v, w)$ is the velocity vector, $\zeta = \nabla \times \mathbf{u}$ the vorticity vector, $\rho(x, y, z)$ the density, $p(x, y, z)$ the pressure, ν the kinematic viscosity, $\triangle$ the Laplacian operator and $F.S.$ denotes the vorticity source from bottom friction. In the following, the vorticity production related to viscous effects will be neglected.

The boundary conditions enable to introduce into equation (6.1) the vorticity contribution coming from the bottom generated turbulence. This additional dispersion term needs to be modeled. As starting point to model the bottom friction, the Newtonian-law for the bottom shear stress τ_b is considered, which reads as follows

$$\frac{\tau_b}{\rho_2} = \nu \frac{\partial u}{\partial z} \tag{6.2}$$

where ρ_2 is the density of the lower layer fluid and $\partial u/\partial z$ the streamwise velocity gradient due to the bottom generated shear stress. Using the definition of the bottom shear stress velocity $u_\star = \sqrt{\tau_b/\rho} = u\sqrt{f_w/8}$, equation (6.2) can be rewritten as follows

$$\nu \frac{\partial u}{\partial z} = \frac{f_w}{8} u^2 \tag{6.3}$$

where $f_w = 4c_f$, with c_f being a quadratic-law friction coefficient. To introduce this term into equation (6.1) the Laplacian operator has to be applied on both hand sides of

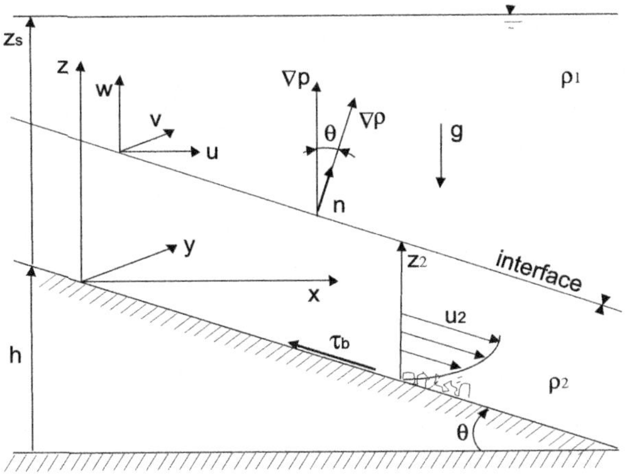

Figure 6.1: Definition diagram for vorticity production in a gradually evolving two-layer flow over downslope.

equation (6.3), leaving to

$$\nu\triangle\left(\frac{\partial u}{\partial z}\right)=\nu\triangle\zeta=\frac{f_w}{4}\left[\left(\frac{\partial u}{\partial z}\right)^2+\frac{\partial^2 u}{\partial z^2}u\right], \tag{6.4}$$

where the streamwise and cross-stream variation of the velocity u has been neglected, since $\partial/\partial y < \partial/\partial x \ll \partial/\partial z$. The term $B.C.$ in equation (6.1) can thus be replaced with the right hand side of equation (6.4) and equation (6.1) reads finally as follows

$$\frac{D\zeta}{Dt}=\zeta\cdot\nabla\mathbf{u}+\frac{\nabla\rho\times\nabla p}{\rho^2}+\frac{f_w}{4}\left[\left(\frac{\partial u}{\partial z}\right)^2+\frac{\partial^2 u}{\partial z^2}u\right] \tag{6.5}$$

It is noted that by integrating the last term on the right hand side of equation (6.5) in the z direction, the bottom boundary condition reduces to equations (6.3) and (6.2).

In the following, the different production terms on the right hand side of equation (6.5) are scaled separately.

The vortex stretching generation of y-vorticity is given by:

$$(\zeta\cdot\nabla\mathbf{u})_y=\left(\frac{\partial u}{\partial z}-\frac{\partial w}{\partial x}\right)\left(\frac{\partial u}{\partial y}+\frac{\partial v}{\partial y}+\frac{\partial w}{\partial y}\right). \tag{6.6}$$

The continuity equation requires that

$$\frac{\partial v}{\partial y}=-\frac{\partial u}{\partial x}-\frac{\partial w}{\partial z} \tag{6.7}$$

Substituting this expression into equation (6.6) it follows

$$(\zeta\cdot\nabla\mathbf{u})_y=\left(\frac{\partial u}{\partial z}-\frac{\partial w}{\partial x}\right)\left(\frac{\partial u}{\partial y}+\frac{\partial w}{\partial y}-\frac{\partial u}{\partial x}-\frac{\partial w}{\partial z}\right). \tag{6.8}$$

Remembering that $\partial/\partial x \ll \partial/\partial z$ and assuming that there is no cross stream variation in u and w, equation (6.8) can be simplified as follows

$$(\zeta \cdot \nabla \mathbf{u})_y = -\frac{\partial u}{\partial z}\frac{\partial w}{\partial z}. \tag{6.9}$$

To evaluate the term $\partial w/\partial z$ we note that the variation of the velocity w in the vertical direction is expected to be proportional to the variation of the water depth above the sill z_s in the streamwise direction, i.e. $\partial w/\partial z \propto u/z_s dz_s/dx$, where z_s is the water depth above the slope. The vortex stretching term can then be written as

$$(\zeta \cdot \nabla \mathbf{u})_y = \frac{\partial u}{\partial z}\frac{u}{z_s}\frac{dz_s}{dx}. \tag{6.10}$$

Thus, the contribution to the vorticity due to vortex stretching is a result of both the streamwise velocity vertical gradient and the acceleration down the slope. Taking as velocity scale the velocity difference between the upper and lower layer ΔU and as vertical scale the vorticity thickness δ_ν, the vortex stretching term is finally approximated as follows

$$(\zeta \cdot \nabla \mathbf{u})_y = \frac{\Delta U^2}{\delta_\nu}\frac{1}{z_s}\sin\theta. \tag{6.11}$$

where θ denotes the angle of inclination of the slope with respect to the longitudinal direction x and the equality $dz_s/dx = \sin\theta$ has been used.

The baroclinic production of y-vorticity reads

$$\frac{\nabla\rho \times \nabla p}{\rho^2} = \frac{1}{\rho^2}\left(\frac{\partial \rho}{\partial x}\frac{\partial p}{\partial z} - \frac{\partial \rho}{\partial z}\frac{\partial p}{\partial x}\right) \tag{6.12}$$

Assuming the horizontal pressure gradient is negligible and using the Boussinesq approximation for small density variations it follows

$$\frac{\nabla\rho \times \nabla p}{\rho^2} = \frac{g}{\overline{\rho}}\frac{\partial \rho}{\partial x} = \frac{g}{\overline{\rho}}\frac{\partial \rho}{\partial n} sin\theta \tag{6.13}$$

with $\overline{\rho}$ being the mean density in the layer. The relation $\partial\rho/\partial x = \partial\rho/\partial n sin\theta$ derives from the flow geometry, in which n is the direction of density gradient and θ denotes the angle of inclination of the interface with respect to the longitudinal direction x (see figure 6.1). For a dimensional analysis the vertical length scale δ_ρ represents the shear layer thickness. Assuming that the interface is approximately parallel to the sill slope, $sin\theta \sim dz_2/dx$, and equation (6.13) can be approximated as follows

$$\frac{\nabla\rho \times \nabla p}{\rho^2} = \frac{g'}{\delta_\rho}\frac{dz_2}{dx}. \tag{6.14}$$

Taking as velocity scale the velocity in the lower layer u_2, the contribution of y-vorticity production due to the bottom roughness can be evaluated as

$$\frac{f_w}{4}\left[\left(\frac{\partial u}{\partial z}\right)^2 + \frac{\partial^2 u}{\partial z^2}u\right] \sim \frac{f_w}{4}\left[\left(\frac{u_2}{\delta_b}\right)^2 + \frac{1}{\delta_b}\frac{u_2}{\delta_b}u_2\right] \tag{6.15}$$

where $\delta_b = z_2/2$ is the vertical scale for the bottom velocity gradient, with z_2 and u_2 being the water depth and velocity in the lower layer, respectively (cf. figure 6.1). Assuming that the vorticity and density gradients are constant across the interfaces, integrating equations (6.14) and (6.15) to obtain the total vorticity produced at a given x-location, amounts to multiplying by δ_ν, δ_ρ and δ_b respectively. The following interaction parameter can be defined

$$\frac{\text{Baroclinic production}}{\text{Vortex stretching}} = \mathcal{Z}_v \equiv \frac{\left|\delta_\rho \left(\frac{\nabla\rho\times\nabla p}{\rho^2}\right)_y\right|}{|\delta_\nu(\zeta\cdot\nabla\mathbf{u})_y|} = \frac{g' z_s}{(\Delta U)^2 \sin\theta}\frac{dz_2}{dx} = F_z^{-2}, \quad (6.16)$$

$$\frac{\text{Friction production}}{\text{Baroclinic production}} = \mathcal{Z}_r \equiv \frac{\left|\delta_b\left(\frac{f_w}{8}\triangle(u^2)\right)\right|}{\left|\delta_\rho \left(\frac{\nabla\rho\times\nabla p}{\rho^2}\right)_y\right|} = \frac{f_w u_2^2}{2\delta_b g' \frac{dz_2}{dx}} = F_\ell^2\left(\frac{dz_2}{dx}\right)^{-1} \quad (6.17)$$

$$\frac{\text{Friction production}}{\text{Vortex stretching}} = \mathcal{Z}_{rv} \equiv \frac{\left|\delta_b\left(\frac{f_w}{8}\triangle(u^2)\right)\right|}{|\delta_\nu(\zeta\cdot\nabla\mathbf{u})_y|} = \frac{z_s}{\ell\sin\theta}\left(\frac{u_2}{\Delta U}\right)^2, \quad (6.18)$$

where $F_z^2 = (\Delta U)^2/(g' z_s)$ is a Froude number defined using the velocity scale ΔU and the vertical length scale z_s and can be understood as the ratio between the limiting effects of buoyancy and of the deceleration of the flow due to the increasing water depth. The length scale $\ell = 2\delta_b/f_w$ represents a "roughness scale" and $F_\ell^2 = |u_2|^2/(g'\ell)$ is a "frictional Froude number", which can be reinterpreted as a measure of the limiting effects of buoyancy on the velocity acceleration (hindered by the friction) in the lower layer. The parameter $\mathcal{Z}_{rv}$ expresses the ratio between the frictional velocity scale u_2 to the baroclinic velocity scale ΔU and the frictional length scale ℓ relative to the total water depth above the slope z_s.

Equation (6.5) tells that the vorticity production due to vortex stretching is balanced by the the stabilizing vorticity contribution related to the stratification and to bottom friction. Moreover, while the vorticity generated by the stratification as well as by the vortex stretching are "intrinsic" contributions related directly to the fluid properties, the term resulting from the bottom friction represents an "external" source of vorticity and is built from the boundary conditions. In this sense, the interaction parameters containing this contribute are the most interesting parameters.

Because there are three parameters defined using three vorticity contributions, they are related to each other and one of them is redundant, i.e. $\mathcal{Z}_v = \mathcal{Z}_{rv}/\mathcal{Z}_r$.

Exp.	g' cm/s^2	Bottom configuration	Mean height	Mean spacing	Equivalent sill height $h'H$ [cm]	Roughness Coefficient f_w
1	4.47	smooth	-	-	20.0	0.01
2	3.97	dense	~ 1.6 cm	0.1 cm	20.9	0.05
3	4.10	sparse	~ 3.5 cm	3.0 cm	20.4	0.07

Table 6.1: Details of the performed experiments. g' is the reduced gravitational acceleration, h' is the equivalent sill height calculated from equation (6.19). In all the experiments the inclination of the sill was roughly $10°$ and the acquisition frequency was 10 Hz.

6.2 Experimental program

Table 6.1 lists the experimental conditions. In all the performed experiments a water depth of $H = 40\ cm$ was used with a constant density difference of roughly 0.4% yielding g' of the order of $4\ cm/s^2$. Experiment 1 was conducted using a smooth bottom, while experiments 2 and 3 have been performed with two different bottom roughness: in experiment 2 artificial turf was used (see figure 6.2 (a)), with a dense close spacing of fine, short roughness elements. Hereby, the mean height of the roughness elements was approximately 1.6 cm with an averaged spacing of 1 mm (cf. table 6.1). The bottom roughness here appears to be the k-type (Perry and Joubert [1969]), the height of the roughness elements playing a role in determining the eddy scale of the turbulence.

In experiment 3 screw anchors were used (see figure 6.2 (b)), with sparse spacing of coarse, tall roughness elements. In this bottom configuration, the mean height of the roughness elements was 3.5 cm with an average spacing of 3 cm. The bottom roughness here appears to be the d-type (Perry and Joubert [1969]), where due to the sparse configuration, the roughness elements act more as single disturbance bodies, with wake turbulence generated in the wake behind the elements and thus, with a less effectiveness in disturbing the overlying flow.

The vertical and horizontal length scales are given by the total water depth H and the sill length L on the right hand side of the sill crest, respectively. The velocity scale is given by $\sqrt{g'H}$ and the flow rate scale by $H\sqrt{g'H}$. The overall Reynolds number $Re = \sqrt{g'H}H/\nu$ is approximately 50,000, with ν being the kinematic viscosity of fresh water.

6.3 Visual observations and qualitative description of the flow

The flow development at the interface for the three experiments is presented in figure 6.3 by a series of instantaneous pictures with a time lag of 2 seconds taken during the maximal exchange regime. In the smooth case (experiment 1) the flow is characterized by the generation of two-dimensional structures, generated by oscillations of the flow rate due to the superimposed barotropic flow (see chapter 5) leading to large billow-like structures at the interface between the two layers. The interface fronts passes right to the sill crest ((a), first and second picture) and the flow in the lower layer begins to accelerate down the sill. In the third and fourth picture, the growth of the large structure can be clearly observed, while in picture five, it starts to collapse. During the collapsing process, small KH billows are generated at the sharp interface created directly in the wake of the large structure. In this smooth case, the wake behind the surge-like flow can fully develop leaving in this region significant entrainment. The rolling-up of the structure followed by a collapsing mechanism in the wake of the structure, forces the depth of the lower layer to reduce its height significantly (in the order of 10% of the total water depth) and this creates the immediately following sharp interface characterized with KH billows. This behavior is periodic and the phenomenon starts again from the beginning (picture 8) (see chapter 5). Experiment 2, with dense bottom roughness shows a different behavior: the increase in the water depth of the lower layer is still present at the interface as a result of the pulsating barotropic flow. But the growth and subsequent roll-up mechanisms of the structures can not be observed anymore, because the turbulence produced by the bottom roughness takes their kinetic energy and redistributes it on smaller three-dimensional scales, favoring the eddy stretching. Without roll-up of the two-dimensional structures, no wake zone is generated. The turbulent fluctuations are increased in this experiment as is discussed more in detail in section 6.5 further below. Experiment 3 with a sparse configuration of bottom roughness lies in between the behavior of the smooth and the dense roughness case: it behaves like the smooth case in the initial accelerating region near the sill crest (pictures 1 to 4), where a growing two-dimensional structure can be observed. However, in this case, the collapse mechanism does not occur and therefore, no wake region is generated in the lee of the surges. This is due to the fact that in this configuration, the sparse distributed bottom roughness, with tall roughness elements, inhibits the roll-up mechanism of the large-scale structures when the structures reach a certain size, and interact with the roughness elements. The roughness stops the counterclockwise rotation of the structure induced by the exchange flow and the two-dimensional turbulence decays very fast with an initial longitudinal vortex stretching followed by the usual three-dimensional vortex stretching.

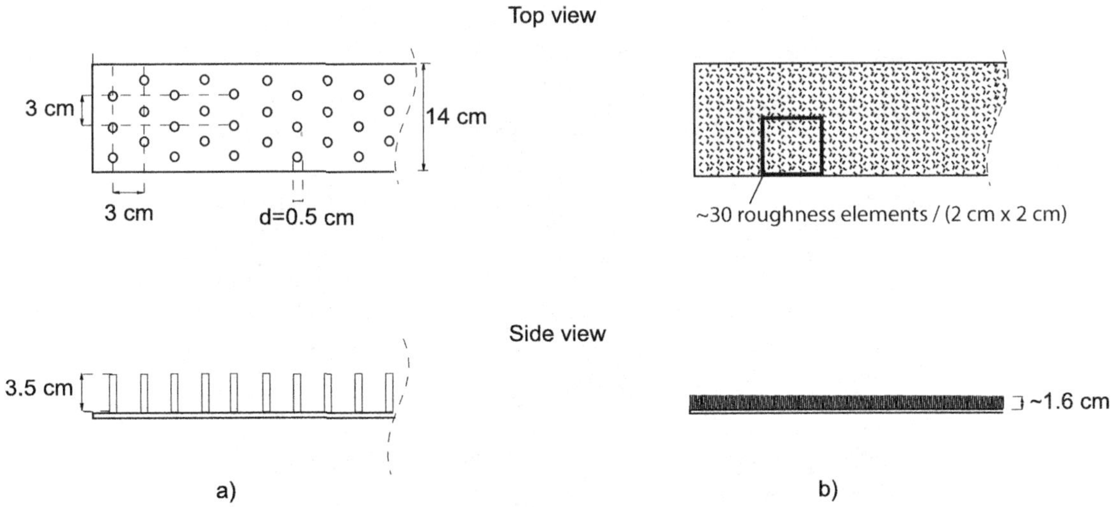

Figure 6.2: Sketch of the roughness used in the rough experiments. a) Experiment 2 with artificial turf, with a dense close spacing of fine, short roughness elements. Hereby, the mean height of the roughness elements was approximately 1.6 *cm* with an averaged spacing of 1 *mm* (cf. table 6.1). The bottom roughness here appears to be the k-type (Perry and Joubert [1969]). (b) In experiment 3 screw anchors were used, with sparse spacing of coarse, tall roughness elements. In this bottom configuration, the mean height of the roughness elements was 3.5 *cm* with an average spacing of 3 *cm*. The bottom roughness here appears to be the d-type (Perry and Joubert [1969]), where due to the sparse configuration, the roughness elements act more as single disturbance bodies.

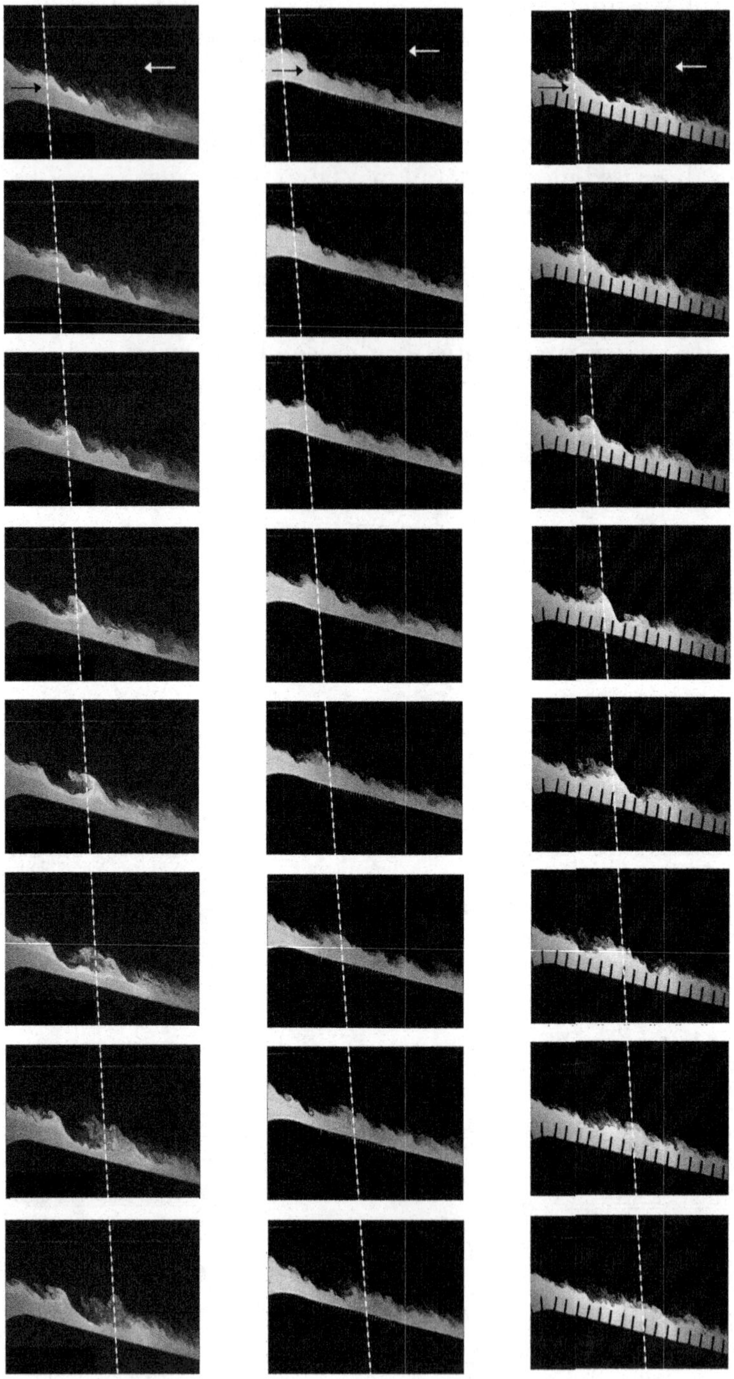

(a) (b) (c)

Figure 6.3: Instantaneous images of experiment 1 (a) the dense roughness case (b) and the sparse roughness case (c) with dye (Rhodamine 6G) mixed in the lower layer (salt water). The time lag between the images is 2 seconds. The image size is 55 *cm* (vertical) x 70 *cm* (horizontal). Dashed lines follow the development of the large-scale interfacial structures.

6.4 Estimation of bottom roughness coefficients and comparison to hydraulic theory results

In this section, results from the PIV measurements are compared to the results from the inviscid hydraulic theory. Quantities in the upper layer are denoted with the index 1 while quantities in the lower layer with the index 2. The bottom roughness coefficients were estimated using the measured flow rates in the upper layer for all three experiments at positions near the sill crest, where the entrainment between the two layers can be neglected (Morin and Loewen [2004]). We expect the flow rates to be decreased in the rough cases, as the bottom roughness effectively results in an increased height of the sill. The normalized flow rates in the upper layer of the rough experiments were thus compared to that of the smooth experiment to obtain the ratio $q_{1c,r}/q_{1c,s}$, where $q_{1c,r}$ and $q_{1c,s}$ are, respectively, the flow rates in the upper layer at the sill crest for the rough (index r) and for the smooth (index s) experiments. The ratio gives a value of 93% for the dense roughness case and 97% for the sparse roughness case as shown in figure 6.4. These values can be used to calculate with the help of internal hydraulic theory an equivalent sill height for the rough experiments. For this purpose, following Zhu [2002], the critical condition at the sill crest for the internal Froude number G, i.e. $G^2 = q^2/g'(z_1^{-3} + z_2^{-3}) = 1$ and the condition $h + z_1 + z_2 = 1$ were used. Herein, h denotes the normalized height of the sill, z_1 and z_2 the normalized water depth in the upper and lower layer, respectively, and $g' = (\rho_2 - \rho_1)/\rho_1 g$, with ρ_1 and ρ_2 being the density of fresh water and salt water, respectively. Assuming that at the sill crest $z_1 = z_2$ and equalizing the critical condition for G for both the smooth and rough experiments the following expression can be obtained

$$h' = 1 - \left(\frac{q_{1c,r}}{q_{1c,s}} \right)^{2/3} (1 - h) \qquad (6.19)$$

where h' denotes the equivalent sill height. Using the estimations as shown in figure 6.4 the equivalent sill heights are $h'_2 = 0.524$ and $h'_3 = 0.510$ for experiments 2 and 3, respectively. With the water depth of 40 cm used in all the performed experiments, this corresponds to heights of 20.96 cm and 20.4 cm for the dense and sparse configuration, respectively (see table 6.1). The internal hydraulic energy was defined as $E = h + z_2 + q^2/g'(1/z_2^2 - 1/z_1^2)$ and calculated using both the measured flow rate and interface position and is plotted in figure 6.5 (a), with the continuous line representing the inviscid hydrostatic prediction of $E = 0.75$, which can be calculated following the method outlined in Zhu [2002]. While the internal hydraulic predictions are in very good agreement for experiment 1, it overestimates the internal energy for experiments 2 and 3 on the right of the sill crest, from $x/L = 0.1$ onward. In this experiments energy is obviously dissipated due to frictional effects and this is the reason for the poor matching between experimental data and theoretical predictions. Zhu and Lawrence [2000] gave a relation to estimate the energy loss in longitudinal direction due to wall and interfacial friction. Considering only the contribution given by the bottom wall, an estimation of its friction coefficient

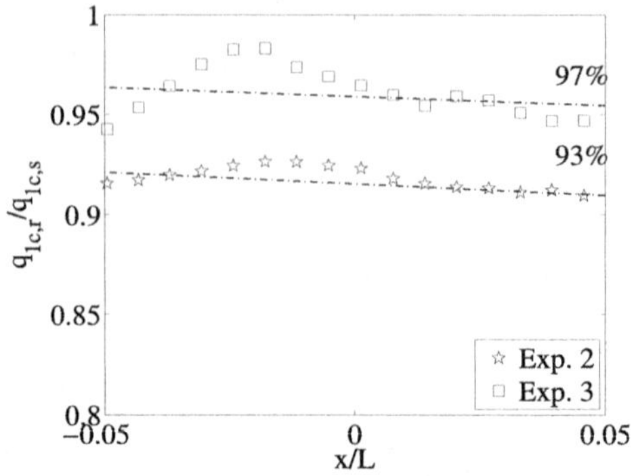

Figure 6.4: Normalized flow rate in the upper layer for the rough experiments compared to the normalized flow rate in the upper layer of the smooth experiment, showing decreased values for the rough experiments.

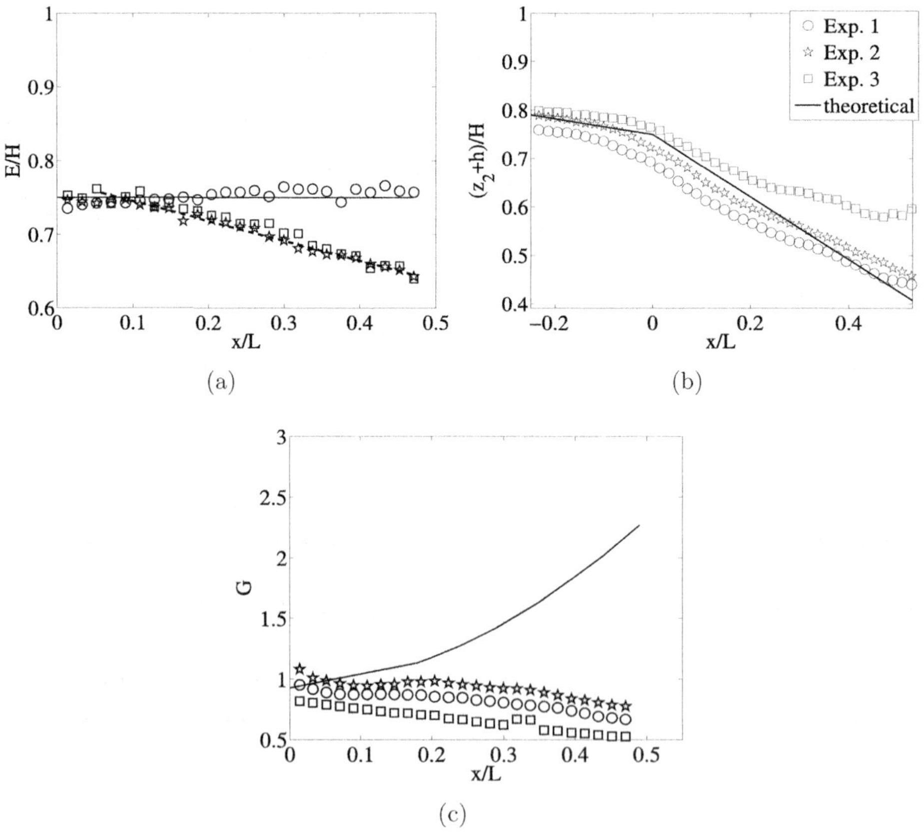

Figure 6.5: Comparison between experimental data (symbols) and theoretical prediction using the hydraulic inviscid hydrostatic prediction (continuous line). (a) Total internal hydraulic energy E, (b) mean interface position and (c) internal Froude number G vs. x.

can be made using the rearranged relation given in Zhu and Lawrence [2000]:

$$f_w = \frac{2}{u_2^2} z_2 \left[S_f \sqrt{H/L_c} - \tau_I \frac{(z_1 + z_2)}{z_1 z_2} \right] \tag{6.20}$$

where S_f is the measured slope of the energy decrease in longitudinal direction, L_c is the length of the connecting channel and $\tau_I = \sqrt{-\rho \overline{u'w'}}$ is the measured interfacial shear stress between the two layers. For the two roughness experiments friction coefficients $f_w = 0.05$ and $f_w = 0.07$ are obtained for the dense and sparse roughness configuration, respectively.

Figure 6.5 (b) shows the position of the averaged interface, along with the predictions of the inviscid hydrostatic internal hydraulic theory (see Zhu [2002]). For the experiments with roughness, the position is higher than the predicted values on the right hand side of the sill crest. This is also consistent with the statement that one effect of the bottom roughness is to increase the height of the sill.

Figure 6.5 (c) shows the densimetric Froude number G, calculated using the predicted flow rate and the measured interface position and using the equivalent sill height for the rough experiments. For all the experiments the theoretical prediction (solid line) over-estimates the measured values on the right hand side of the sill crest, which is probably due to the interface position being overestimated.

6.5 Mean flow characteristics

In figure 6.6, the averaged velocity fields (left column) and the Reynolds-stress $\overline{u'w'}$ distribution (right column) are shown for the three performed experiments. A first noticeable difference between the smooth experiment and the rough experiments is the decrease, for rough bottoms, in the velocity difference between the two layers, due to the reduced acceleration down the slope as compared to the smooth case, and also a decrease of the shear layer thickness. The Reynolds stresses are significantly higher in the two rough experiments with a thinner sheared layer.

The vorticity and density thicknesses, first mentioned in section 6.1, were obtained from the inverse slope of the best-fit line through the normalized velocity profile between the 15% and 85% values (Pawlak and Armi [2000]):

$$\delta_\nu = \left(\overline{\frac{d\overline{u}}{dz} |_{\overline{u}=0.15}^{\overline{u}=0.85}} \right)^{-1} \quad \text{and} \quad \delta_\rho = \left(\overline{\frac{d\overline{\rho}}{dz} |_{\overline{\rho}=0.15}^{\overline{\rho}=0.85}} \right)^{-1},$$

respectively, where $\overline{u}$ and $\overline{\rho}$ are the normalized velocity and density, ranging from 0 to 1. These parameters are plotted for all the experiments in figure 6.7.

The vorticity thickness shows a faster increase for the smooth case (see figure 6.7 (a)), followed by the dense roughness case and, finally, the sparse roughness case. The mixed layer thickness (figure 6.7 (b)) is larger right to the sill crest for the smooth case while it shows smaller values for the rough cases. The wake behind the pulsating surge-like currents observed for the smooth experiment can not develop right to the sill crest: the bottom roughness acts as a catalyst in the energy cascade process characterized by energy transfer from larger to smaller scales and accelerates the dissipation process. This results

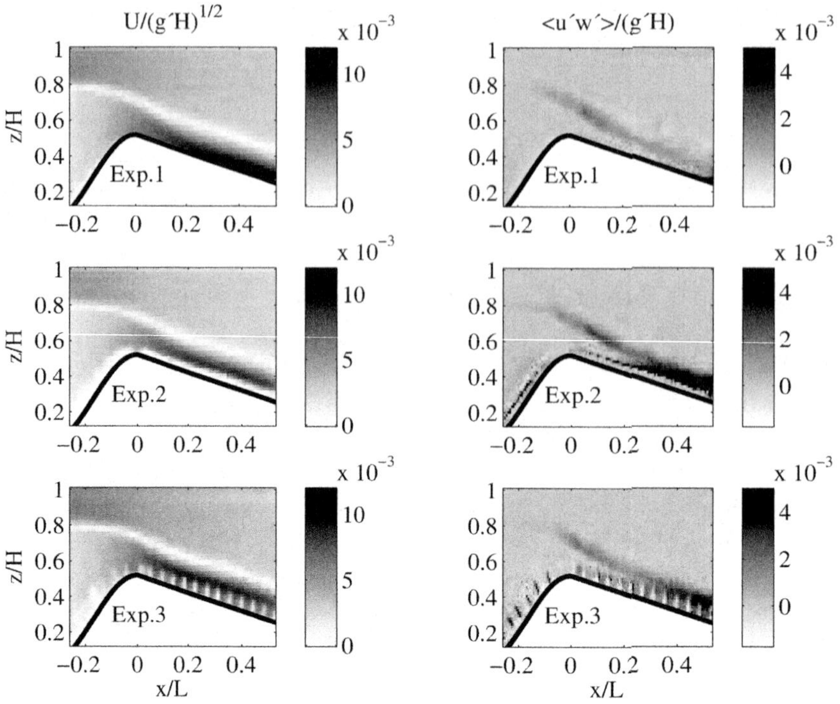

Figure 6.6: Mean flow characteristics in the measured flow field. (a) Normalized averaged velocity field (left column) and normalized Reynolds stresses $\overline{u'w'}$ distribution (right column) for all the experiments.

in a reduced increase of the shear and density layer thickness downstream of the sill crest.

In Pawlak and Armi [1998] the effects of acceleration on the initial region right to the sill crest of the developing shear layer were parameterized using an acceleration parameter, defined as $T_A = (\delta/u_2)(du_2/dx)$, in which δ is the shear layer thickness and u_2 is the representative velocity for the lower layer. This ratio can be interpreted as a ratio of an eddy time scale to an acceleration time scale. T_A is plotted in figure 6.7 (d) for all the performed experiments. Left to the sill crest, they all have similar values. For $x/L > 0$ the acceleration parameter for the smooth case (experiment 1) is higher than in the rough cases (experiments 2 and 3). This means that the time scale for the acceleration is smaller or that the eddy time scale is higher for the smooth case compared to the rough cases, thus indicating a lower acceleration down the sill, and so a lower input of energy due to the sill slope in the rough experiments as compared to the smooth experiment.

The effect of the roughness can also be noted on the density layer properties. The normalized concentration profiles are plotted in figure 6.8 (a): the concentration profiles becomes sharper when the roughness is increasing and thus, the interface in the rough cases is more stable. This is also demonstrated in the other pictures of figure 6.8, where the concentration at the density interface is compared to the mean concentration in the density layer for experiment 1 (figure 6.8 (b)), for experiment 2 (figure 6.8 (c)) and for experiment 3 (figure 6.8 (d)). As the roughness increases, the gap between the data in-

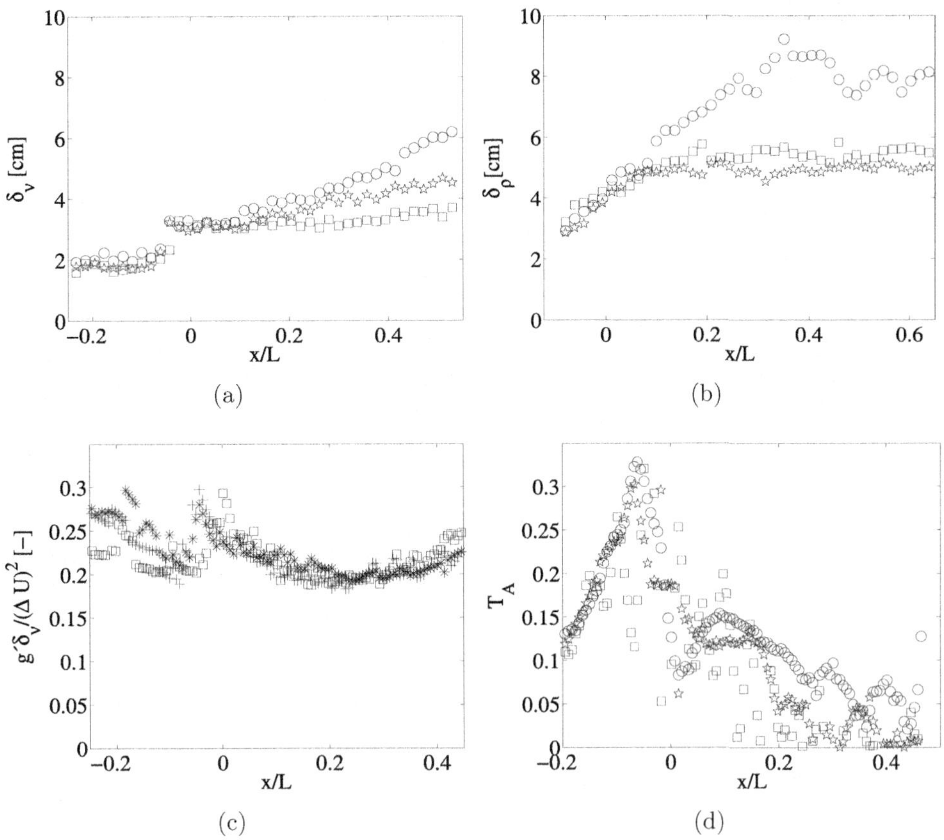

Figure 6.7: Time averaged quantities for the performed experiments. (a) Vorticity thickness and (b) density layer thickness, bulk Richardson number (c) and acceleration parameter T_a (d) vs. x.

creases remarkably. This is somehow contradictory, because bottom roughness normally increases the mixing, and thus makes the density profile less sharp as compared to a smooth case. However, in the flow treated in this study, there are two main sources for mixing between the upper and lower layer: the first one is given by the wakes developed by the large-scale structures periodically generated at the interface and triggered by the barotropic oscillation, while the second one is given by the small-scale, bottom generated turbulence. In the smooth experiment the main source of mixing is given by the wake region developed in the collapsing phase of the large-scale surges mechanism. In experiment 2, the main source of mixing is given by the bottom generated turbulence and the contribute coming from the large-scale surges is unimportant. In experiment 3, the bottom roughness inhibits the collapsing phase of the large-scale barotropic surges and consequently, the wake region in which mixing normally takes place. Moreover, the bottom generated turbulence interacts less intensively with the interface in this case as compared to experiment 2. Both these effects result in reduced mixing as compared to the smooth experiment. Thus, the contribution to the mixing at the interface given by the two-dimensional, large-scale structures seems to prevail as compared to the contri-

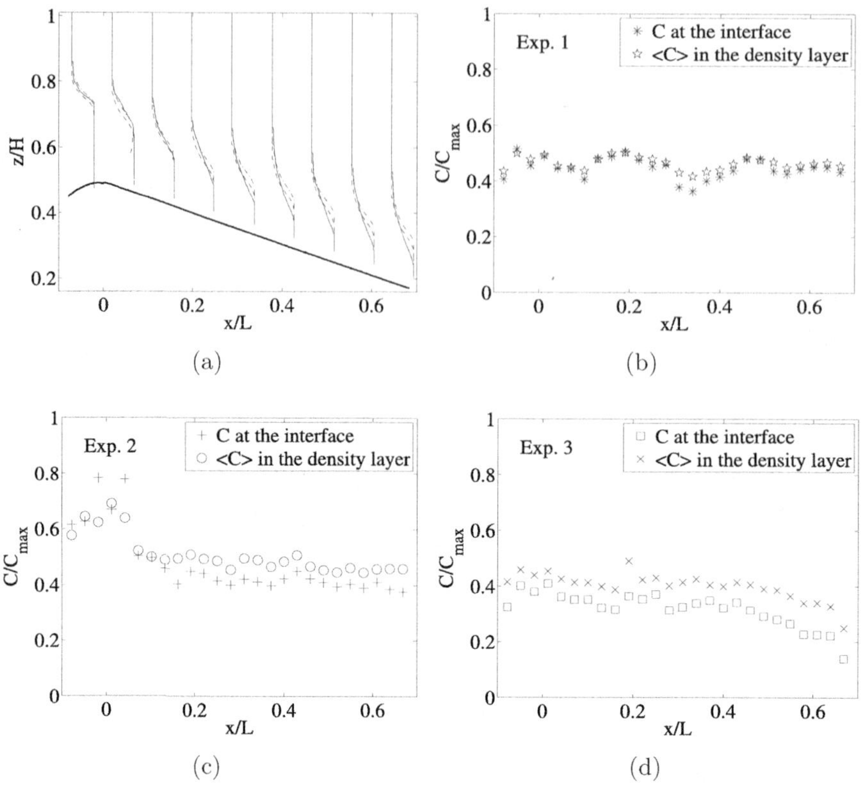

Figure 6.8: Averaged concentration values. (a) Mean concentration profiles for the smooth case (experiment 1, continuous line), the dense roughness case (experiment 2, dashed line) and the sparse roughness case (experiment 3, dash-dot line). (b) Comparison between the concentration at the interface and the mean concentration in the density layer (15%-85% bounded) for the smooth case, the dense roughness case (c) and the sparse roughness case (d). The higher gap between the concentration values for the roughness experiments suggests that the density profiles has been modified by stretching them.

bution given by the bottom roughness, resulting in more sharp density interfaces in the rough experiments. Finally, we point out that, in fact, if one looks at the instantaneous density profiles, they show sharper density interfaces in the smooth case as compared to the rough cases, but in the average the results show the contrary.

Also interesting to note is that the concentration at the density interface is decreasing from a value of roughly 0.45 for experiment 1 to a value of 0.4 for experiment 2 and finally to a value of 0.3 for experiment 3. This could suggests that the mixing in the rough cases is not so efficient as in the smooth case. This contraddictory behavior will be further discussed in section 6.8.

6.6 The effect of bottom roughness on the interfacial large-scale structures

The difference between the two experiments with roughness is that the configuration with dense close spacing and fine short roughness elements is more effective in damping the formation of the two-dimensional structures at the interface because it inhibits the acceleration down the sill of the lower layer and causes a quicker energy transfer from the large to the small turbulence scales. Sparse spacing of coarse, tall roughness elements act almost as single disturbance bodies and the turbulence produced between the elements is not interacting so intensively with the interface as compared to the dense configuration.

Due to the larger height of the roughness elements in the sparse roughness experiment the growth process of the two-dimensional structures generated at the interface is disturbed and the collapse of these structures is hindered, so that no wake zone can be observed behind them. In contrast to free shear layers which grow continuously downstream, shear layers generated by submerged roughness elements grow only to a finite thickness Ghisalberti and Nepf [2004]. Because these shear layers are characterized by coherent vortex structures and rapid vertical mixing, their thickness controls the exchange between the roughness zone and the overlaying water. In the dense roughness case, the velocity gradients are much stronger than in the sparse roughness case: in figure 6.9 (a) the characteristic velocity profile for a dense roughness case is depicted schematically. The measured averaged velocity profiles for this case are shown in figure 6.10 (a) and it can be noted that the velocity is zero near the bottom of the sill due to the dense distribution of the roughness elements, but then it increases very fast (cf. also figure 6.6). In the sparse roughness case (schematically depicted in figure 6.9 (b)) a more gentle increase of the velocity in the lower layer occurs to reach similar velocities as in the dense roughness case. This can also be seen in the measured averaged velocity profiles for the sparse roughness case in figure 6.10 (b)), where the velocity is low between the roughness elements and increases slowly in vertical direction. Also the mixing layer is growing faster and becomes larger in the dense roughness case than in the sparse roughness case (cf. figure 6.7): this suggests that the interaction between the boundary layer and the velocity shear interface between the two layers is more intensive for the dense roughness case than for the sparse roughness case and the two-dimensional turbulence which would be generated at the interface shear layer due to the superimposed barotropic flow decays much more faster in three-dimensional turbulence in the dense roughness case as compared to the sparse roughness case.

In the sense of Perry and Joubert [1969], the dense, short roughness used in the dense roughness case is more of the *d*-type, with a higher interaction of the bottom generated turbulence with the interface between the two layers, while the sparse, tall roughness employed in the sparse roughness case are more of the *k*-type, where there is a generation of a turbulent shear layer between the roughness elements, but this does not interact with the sheared interface (Thorpe [1983]). Thus, in the sparse roughness case, the turbulent bottom layer spreads vertically less rapidly than the two-dimensional structures generated at the interface due to baroclinic/barotropic effects.

Figure 6.11 shows the time series of the velocity shear interface, which is here defined as $(h|_{\overline{u}=0.85} + h|_{\overline{u}=0.15})/2$, at three different longitudinal sections ($x/L = -0.05$, $x/L =$

0.19, $x/L = 0.42$) for the smooth case (a), the dense roughness case (b) and the sparse roughness case (c). While for the smooth case the mean interface position is decreasing as $\sim 0.7 \rightarrow 0.6 \rightarrow 0.55$ from the sill crest to the right, for the rough cases it starts with a mean value of ~ 0.75 decreasing to ~ 0.65 and reaching finally ~ 0.55. This means that in the rough cases the depth of the lower layer is higher at least in the first accelerating region compared to the smooth case. In the dense roughness case has the smallest amplitude of the sheared interface fluctuations while in the smooth and the sparse roughness cases, it has the same order of magnitude, at least for the sections on the right hand side of the sill crest, but in experiment 3 the frequency is higher. This higher frequency is again the result of the shorter time life of the internal waves, which collapse faster compared to the smooth case.

Figure 6.12 shows the interface fluctuation record for a time series of 400 s duration, constructed from a vertical cut at a fixed location in the spatial image, i.e. at $x/L = 0.24$. A spatial length scale can be inferred by assuming a convective velocity of $\Delta U/2$. The vertical fluctuations of the density interface are visibly reduced for the rough cases as compared to the smooth case. Moreover, the horizontal size of the large-scale structures is visibly reduced in the rough cases as compared to the smooth case.

Figure 6.13 (a) shows the standard deviation of the density interface fluctuations shown in figure 6.12 as a measure of the growth rate of the large-scale two-dimensional structures. It is worth noting that while for the smooth experiment there is a constant increase of the standard deviation as the flow moves down the slope, the rough experiments show an opposite behavior, with the standard deviation decreasing in longitudinal direction. In the sparse roughness case there is an initial increase in the region $0 < x/L < 0.15$, and then it continuously decreases, indicating an initial growth rate, but a rapid collapse of the large-scale structures at the interface. In the dense roughness case there is a constant decrease, suggesting that no two-dimensional structures are generated at the interface. Moreover, the standard deviation for the rough experiments is higher as compared to that of the smooth case in the initial developing region $x/L < 0.3$, while in the growing region of the large scale structures, the fluctuations of the density interface become larger

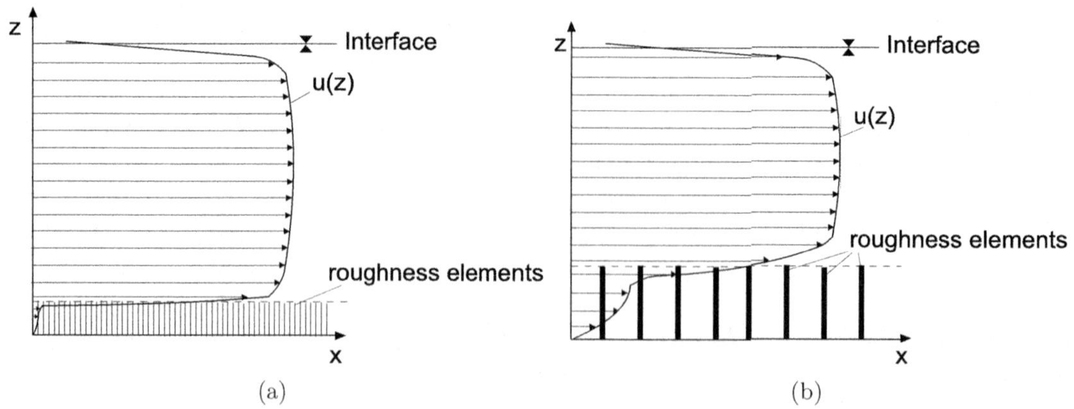

Figure 6.9: Schematic sketch of the velocity profiles with enhanced bottom roughness. (a) Velocity profiles for dense close spacing short roughness elements and (b) for sparse coarse tall roughness elements.

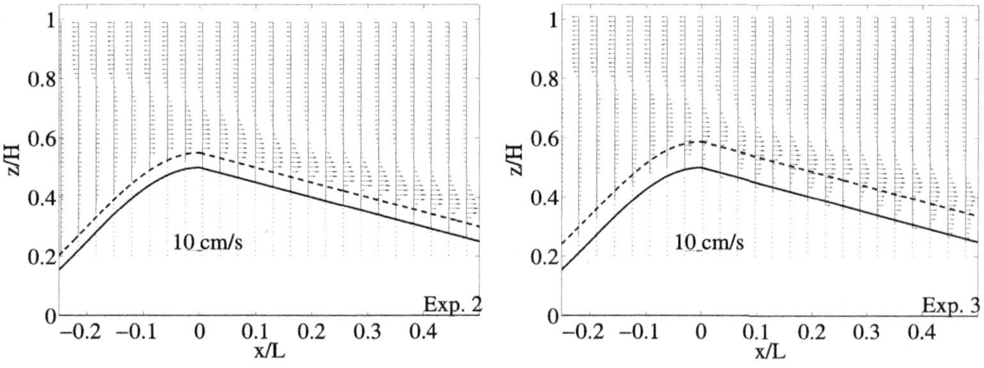

(a) (b)

Figure 6.10: Measured averaged velocity distributions for (a) the dense roughness case and (b) sparse roughness case.

than those of the rough experiments. This is also consistent with the visual observations described above.

Figure 6.13 (b) show an estimation of the horizontal (empty symbols) and vertical (filled symbols) length scales of the large, two-dimensional structures at the interface, calculated using the time scale T_I obtained from the autocorrelation functions of the horizontal and vertical velocity fluctuations and the mean velocity shear $\Delta U/2$ as convective velocity scale. For the dense roughness case, both the horizontal and vertical sizes of the large-scale structures at the interface are reduced, while for the sparse roughness case only the vertical scale is reduced, as compared to the smooth case, while the horizontal scale is roughly the same. As noticed previously in other studies (Negretti and Jirka [2007], Morin and Loewen [2004]) the period of the oscillations in the net flow rate due to the unsteady external forcing are not exactly the same as those of the interface. This was explained by Negretti and Jirka [2007] as different control mechanisms of these oscillations: the oscillation in the net flow rate is controlled by the translation motion of the fluid through the channel; the fluctuations of the interface, however, are the result of two mechanisms, namely surge-like accumulations at the sill crest triggered by the barotropic flow and the additional baroclinic wave growth at the interface (i.e. even for steady exchange flow). Apparent non-linear interaction between these mechanisms causes a shorter interfacial wave period as opposed to the pure barotropic flow fluctuation period. Here the amplitude of fluctuations of the interface and the growth of the surge-like flows is hindered by bottom friction: to assure that the necessary amount of volume for the barotropic equilibrium can be exchanged from one reservoir to the other, the interface can only change its frequency of oscillation, if the amplitude growth is limited by friction. Thus, the frequency of oscillations of the interface, and so the frequency by which the surge-like flows are generated, does depend not only on the geometrical quantities of the two reservoirs and the connecting channel, but also by the amplitude and maximal growth of the large-scale structures. From figure 6.12 (a), we estimate the period of events characterized by the large-scale billows to be approximately 18.0 s for the smooth case and also for the sparse roughness case, while for the dense roughness

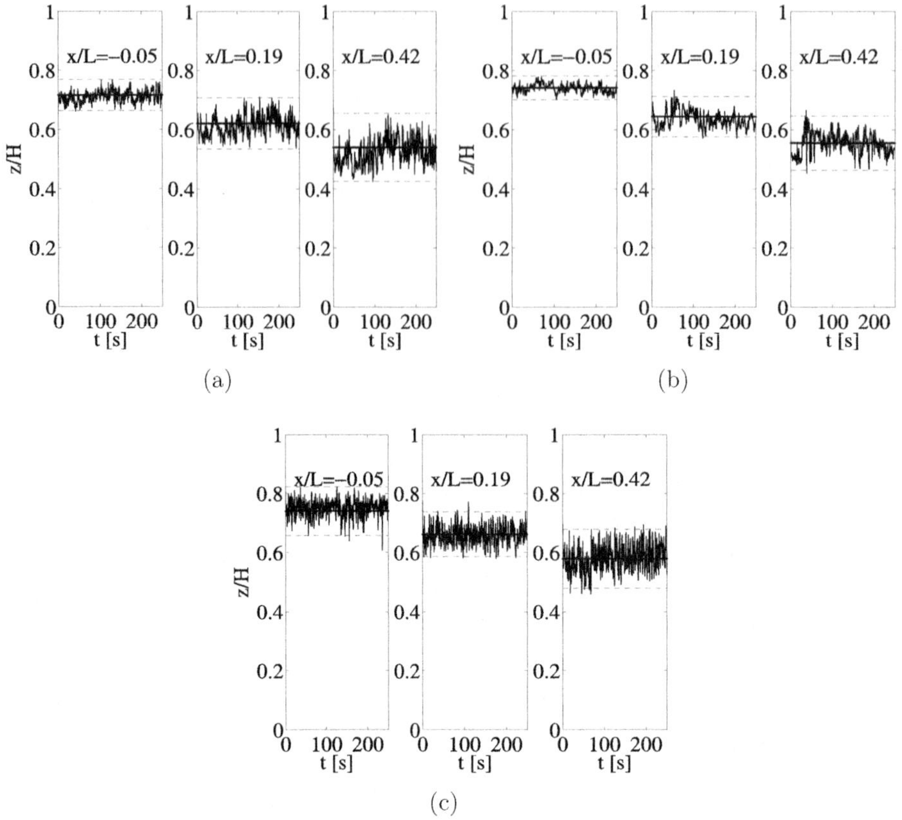

Figure 6.11: Time series of the instantaneous velocity shear layer interface at three different longitudinal positions. (a) experiment 1, (b) experiment 2, (c) experiment 3. The highest amplitude of the fluctuations is observed for experiment 1, followed by experiment 3 and finally for experiment 2.

case it is estimated to be 11.8 s. The dependence of the size of the large scale structures versus the observed periods of generations is reported in figure 6.14 for the three performed experiments and shows that the generation period increases linearly with the increase of the size of the large-scale structures at the interface, with a slope of roughly 0.059. Thus we propose the following relation:

$$\frac{A_I}{H^2} = CT_{obs} \tag{6.21}$$

where $A_I = L_{I,hor}L_{I,ver}$ is a rough estimation of the area of the large-scale structures and T_{obs} is the related observed period of generation. For a more exact estimation of the value of the constant C more experiments are needed as well as a more detailed estimation of the period of generation of these structures, for example through the spectral analysis of the density interface fluctuations. Due to the relatively low image acquisition frequency this will not be done here.

The different behavior between the two rough experiments can be explained, again, by the different intensity of interaction between the bottom generated turbulence and

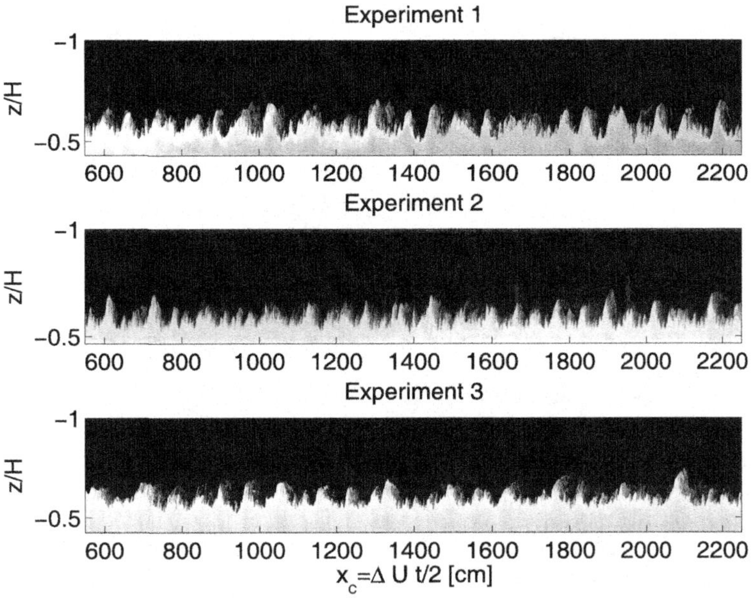

(a)

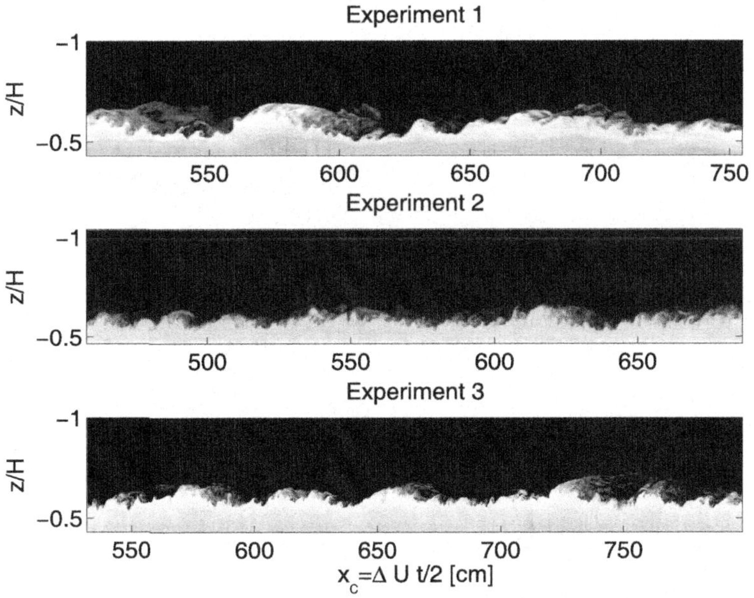

(b)

Figure 6.12: (a) Reconstructed images of the interfacial fluctuations for all the performed experiments at $x/L = 0.24$ vs $x_c = t\Delta U/2$, where $\Delta U/2$ is assumed as mean convective velocity. The values of ΔU are 0.05 m/s, 0.046 m/s and 0.05 m/s are for experiment 1, 2 and 3 respectively. The pictures shows the large internal breaking waves as well as KH billows, with a higher frequency. (b) is a zoom in of the time series showed in (a).

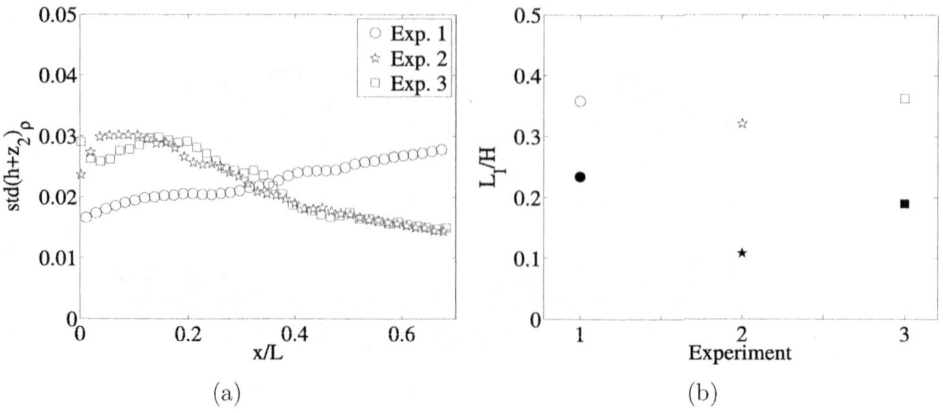

Figure 6.13: (a) Standard deviation of the density interface fluctuations for the three experiments. (b) Estimated horizontal (empty symbols) and vertical (filled symbols) length scales of the large, two-dimensional structures at the interface, calculated using the time scale obtained from the autocorrelation functions of the horizontal and vertical velocity fluctuations and the mean velocity shear as convective velocity scale. Symbols are as given in figure 6.7.

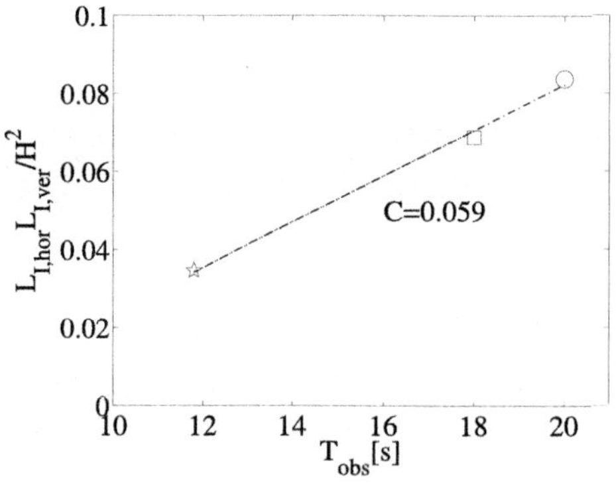

Figure 6.14: Size of the large-scale structures observed at the interface versus the observed period of generation: the higher the size of the structures, the higher the period of generation. The increase seems to be linear.

the interfacial, two-dimensional turbulence: in the dense roughness case the bottom roughness interacts more with the interface and results, as shown previously, in hindering, in an early stage, the development of the large-scale structures at the interface. In the sparse roughness case, on the other hand, there is a limited interaction between the bottom generated turbulence and the interface between the two layers and the large-scale structures can still grow, reaching similar, but still reduced sizes as compared to the smooth case (cf. figure 6.13 (b)). This explains the similar periods observed for the generation of the two-dimensional structures at the interface.

On the basis of these results, we finally conclude that the roughness reduces the vertical and longitudinal size of the two-dimensional structures and that dense roughness is more effective for this scope. Moreover, the period of generation of these large-scale structures

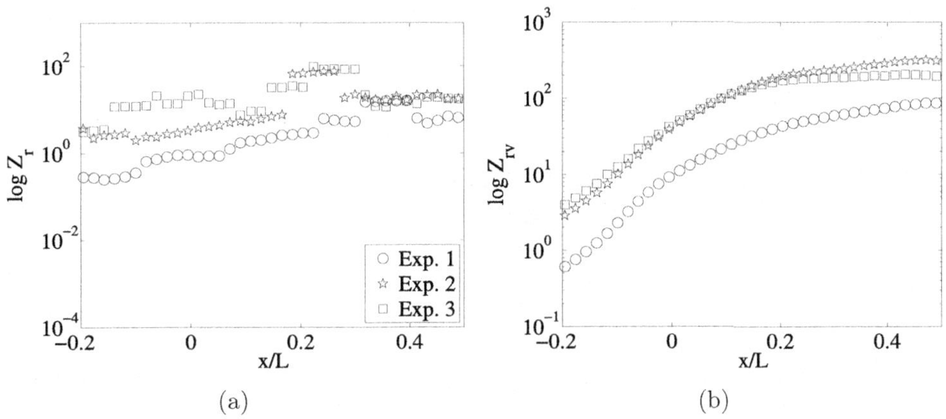

Figure 6.15: Vorticity parameters expressing the ratio between the three main contribution of vorticity production in a two-layer stratified flow down a slope with enhanced bottom roughness. (a) $\mathcal{Z}_r$ as ratio between the friction vorticity production and the baroclinic production due to stratification and (b) $\mathcal{Z}_{rv}$ expressing the ratio between the friction production and the vortex stretching production due to the acceleration down the slope vs. x (cf. equations (6.16, 6.17, 6.18)).

at the interface shows to depend linearly on the size of the large-scale structures.

6.7 The effect of bottom roughness on the turbulence production

The vorticity parameters $\mathcal{Z}_r$ and $\mathcal{Z}_{rv}$ which has been defined in section 6.1 are plotted in figure 6.15.

The vorticity parameter $\mathcal{Z}_r$ (equation (6.17)) is given by the product of the frictional Froude number F_ℓ and the inverse of the inclination of the interface (figure 6.15 (a)). The data for the smooth experiment show a steeper increase for $x/L < 0.3$, and a constant increase after. The data for the two rough experiments show a more gentle increase compared to the increase of the smooth case and the constant phase starts earlier, i.e. at $x/L \sim 0.2$. The steeper trend of the smooth case is because in this case the only restoring force is given by buoyancy, while in the rough cases the gap between frictional and baroclinic production is smaller and they both work for stabilizing the flow. This also explain why the equilibrium is reached earlier by the rough cases as by the smooth case.

The third parameter $\mathcal{Z}_{rv}$, giving the ratio between the roughness vorticity production and the vortex stretching, is plotted in figure 6.15 (b) and it shows to increase very fast in the region $-0.2 < x/L < 0$, just before the sill crest for all the three experiments, with the same rate. The sharpness of the curves decreases then between $0 < x/L < 0.25$ for experiment 1, between $0 < x/L < 0.18$ for the dense roughness case and between $0 < x/L < 0.1$ for the sparse roughness case, but still with the same gradient. This change in the slope of the curves is given by the effects of the acceleration down the sill, so that right to the sill crest, vortex stretching increases faster than the friction production of

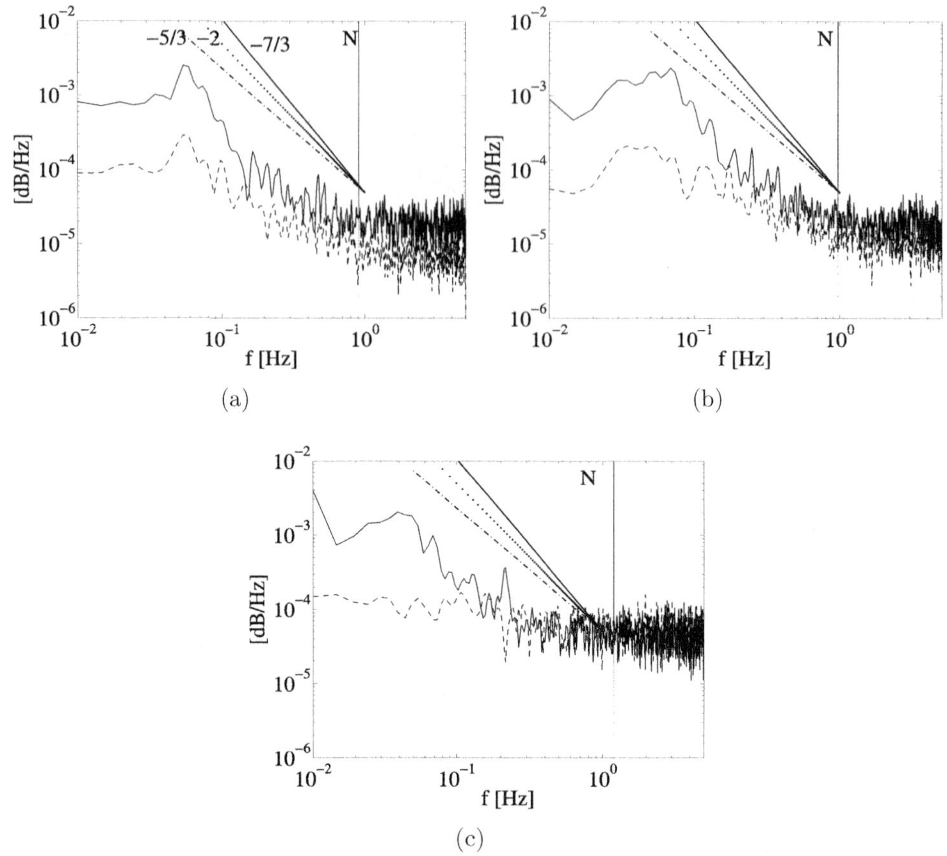

Figure 6.16: Spectral distribution of the velocity fluctuations in streamwise (continuous line) and in vertical direction (dashed line) for experiment experiments 1 (a), 2 (b) and 3 (c). Spectral slopes of $-7/3$, -2 and $-5/3$ are plotted for comparison along with the natural buoyancy frequency N.

vorticity. The third region, for $x/L > 0.25$, $x/L > 0.18$ and $x/L > 0.1$ for experiment 1, 2 and 3 respectively, is characterized by a constant slope for the two rough experiments, as the friction contribution becomes more important, while for the smooth case a reduced slope is present. In conclusion, the behavior of the three vorticity parameters shows that the bottom roughness plays a twofold role in stabilizing the interface. First, it reduces the acceleration down the sill and consequently the velocity in the lower layer. Second, it increases the turbulent energy transfer between the lower layer and the sheared interface, resulting in limiting the growth (in the dense roughness case) or the collapse (in the sparse roughness case) of the large-scale surge-like flows.

These effects can also be observed in the spectral distribution of the fluctuations of the main and vertical velocity components at the interface (see figure 6.16): for the smooth case, the initial part of the spectrum is better described by a $-7/3$-slope and only in the second part by the $-5/3$-slope for the streamwise velocity component (continuous line). For the other two experiments, the spectrum is better described from the beginning by the $-5/3$-slope. The spectral slope is a measure of the rate of energy transfer between

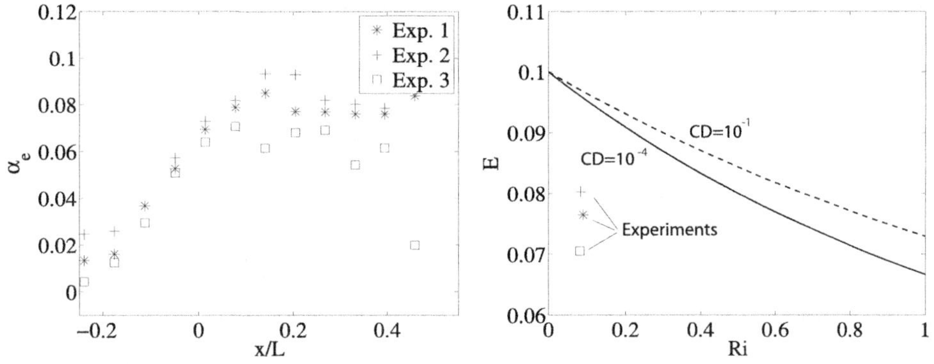

(a) (b)

Figure 6.17: entrainment coefficients of the upper-layer fluid along the longitudinal direction (a) and entrainment coefficients compared with the relation proposed by Dallimore and Ishikawa [2001] vs. the Richardson number Ri (b). Symbols represent averaged entrainment coefficients of the performed experiments, while the curves represent equation (6.23) for different bottom drag coefficients c_D.

the scales since the Reynolds number is large enough. An important issue in the study of stratified flows is to separate the large scale generated frequencies caused for example by internal waves, and the contribution by the buoyancy driven small scales. Low frequency ranges in stratified flows have shown to be better described by -2 and $-7/3$-slopes, for frequencies usually lower than the buoyancy frequency N [Lien and Dairiki, 1998]. Despite the fact that at these scales the limits of the PIV averaging becomes significant, the energy in the last part of the spectrum shows a distinct separation between the longitudinal (continuous line) and vertical (dashed line) velocity fluctuations spectra (see figure 6.16) and they fall together only at higher frequency than the natural buoyancy frequency N for the smooth case, while they fall together very early for experiments 2 and 3, at frequencies which are almost one order of magnitude smaller than their natural buoyancy frequency N. This is an additional indicator that the energy cascade for the rough cases starts earlier compared to the smooth case.

6.8 The effect of bottom roughness on mixing and entrainment

Given the velocity field data, estimates of the entrainment coefficients can be obtained. Following Ellison and Turner [1959], the entrainment coefficient is defined as

$$\alpha_e = \frac{1}{V}\frac{dq}{dx} \quad \text{with} \quad V = \frac{\int u^2 dz}{\int u dz} \tag{6.22}$$

where q is the flow rate per unit width in the layer of interest and V is a representative velocity for the layer. As pointed out in the introduction, many authors tried to give a general relation in order to predict the entrainment coefficients as a function of the Richardson number [Ellison and Turner, 1959, Christodoulou, 1986] and both the

Richardson number and the bottom friction coefficient [Hebbert and Loh, 1979, Kessel and Kranenburg, 1996, Dallimore and Ishikawa, 2001]. Here, we also use the relation proposed by Dallimore and Ishikawa [2001], where the entrainment coefficient is defined as follows

$$E = \frac{C_k C_D^{3/2} + C_s}{Ri + 10(C_k C_D^{3/2} + C_s)} \tag{6.23}$$

where C_k and C_s are coefficients measuring the efficiency of the boundary-introduced turbulent kinetic energy and of the local shear production, respectively. Past experimental data examined by Cherman and Corcos [1978] suggested universal values for $C_k = 2.2$ and $C_s = 0.2$. $Ri = g'h_2/u_2^2$ is the Richardson number as defined also in Ellison and Turner [1959], with h_2 and u_2 being respectively the depth and the characteristic velocity in the lower layer and C_D is the bottom drag coefficient. Figure 6.17 (a) shows the entrainment coefficients for all the performed experiments and the higher values are found in the region $0 < x/L < 0.15$. The overall higher entrainment coefficients are observed for the the dense roughness case, which are of the same order of magnitude of the entrainment coefficients found by Morin and Loewen [2004] and Fouli [2006], approaching the value 0.1. The entrainment coefficients for the smooth case are similar to the entrainment coefficients observed for the dense roughness case: the high turbulence intensities at the interface in the rough experiments favorizes the mixing as the interaction between the bottom boundary layer and the interface is very intensive, as demonstrated by the Reynolds stresses distribution (cf. figure 6.6). In the sparse roughness case the lowest values for the entrainment values are reported (about 30% less as compared to experiments 1 and 2). This reduction is a result of a lower interaction between the bottom boundary layer and the interface between the two layers, on one hand, due to the reduced velocity gradients at the bottom (cf. figure 6.9) and, on the other hand, this results from the fact that the large-scale structures do not develop a wake zone like in the smooth case. In the smooth experiment the main source of entrainment is given by the wake region developed in the collapsing phase of the large-scale surges mechanism. In the dense roughness case, the main source of entrainment is given by the bottom generated turbulence, due to the very strong velocity gradients. Even if the large-scale surges are not observed in this experiment, entrainment coefficients are in the same order as in the smooth case. In the sparse roughness case, the bottom roughness inhibits the collapsing phase of the large-scale barotropic surges and thus, the wake region in which entrainment normally takes place. Moreover, the bottom generated turbulence interacts less intensively with the interface in this case as compared to the dense roughness case. Both these effects result in reduced entrainment coefficients as compared to the other two experiments. These results are also briefly summarized in table 6.2. In figure 6.17 (b), the averaged entrainment quantities reported in all the experiments are plotted (symbols) versus the Richardson number Ri as defined in Ellison and Turner [1959]. The lines represent the prediction proposed by Dallimore and Ishikawa [2001] as in equation (6.23) for different values of the drag coefficient. Increasing the drag coefficient also the entrainment should increase. This is only partially valid for the dense roughness case, as compared to the smooth experiment. For the sparse roughness case the entrainment is lower than in the other two cases, given the same Richardson number. The prediction by Dallimore and Ishikawa [2001] overestimates the entrainment coefficients observed here.

Exp.	growth of surges	collapse of surges	Main source of vorticity production	Main source of mixing
1	yes	yes	surges (barotropic)	surges (barotropic)
2	no	no	bottom roughness	bottom roughness
3	yes	no	baroclinic	surges (barotropic)

Table 6.2: Summary of the results obtained for the three performed experiments relative to the large-scale structures growth and collapse mechanisms, the vorticity production. For damping the large-scale surges, the most effective possibility is represented by the dense roughness case. The primary contribution for mixing is given by the large-scale structures.

6.9 Summary and conclusions

Results of an experimental study on stratified exchange flows down a submerged slope with a superposed barotropic pulsating flow were presented. Two different types of roughness have been used in the experiments and their results have been compared with results relative to a smooth experiment (experiment 1). In the dense roughness case, artificial turf was used as local enhanced bottom roughness over the sill, with a mean height of 1.6 *cm* and with a very high density. In the sparse roughness case, screw anchors were used, with a height of 3.5 *cm* and a reduced density. In such flows, three different mechanisms are responsible for the vorticity production: a first source of vorticity is given by the vortex stretching due to the variation in height of the channel because of the sill placed on the bottom. A second source is represented by the baroclinic generation of vorticity, given by the inclination of the interface. In addition, the enhanced bottom roughness is contributing to the vortex stretching. By dimensional analysis of the vorticity equation, three parameters have been defined, in order to analyze the weight of each contribution for every experiment. The behavior of the three vorticity parameters has shown that the bottom roughness plays two important roles in stabilizing the interface: firstly, it reduces the acceleration down the sill and so, the velocity in the lower layer. Secondly, it increases the turbulent energy transfer between the lower layer and the sheared interface, resulting in hindering the growth (in the dense roughness case) or the collapse (in the sparse roughness case) of large-scale surge-like flows.

Bottom friction coefficients and an equivalent sill height have been estimated through comparison of experimental data with the internal hydraulic theory. The results showed that the roughness shifts the interface position as one effect of roughness is to increase the height of the sill. A remarkable internal energy loss for the rough cases has also been observed.

In the smooth case, the shear and density layer thickness increase faster compared to the rough cases. Averaged velocity fields and Reynolds stresses distributions show a decrease of the velocity difference between the two layers and an increase of the Reynolds stresses in the shear layer ranging from the smooth case to the rough cases. The roughness has shown to make much more sharp the density profiles.

The structure of the flow has been examined in detail, with a physical description of the wave mechanisms at the interface. An estimation of the length scales of the large-scale structures generated at the interface has been made using the autocorrelation functions of the vertical and horizontal velocity components and the results demonstrated the reduction of the size of the structures in the rough experiments. Moreover, the period of generation of these large-scale structures at the interface seems to depend linearly on the size of the large-scale structures.

The different bottom roughness elements have shown to behave differently with respect to the interfacial wave activity: if the bottom roughness is very dense, the velocity gradients are much more stronger as compared to sparser roughness. The interaction between the boundary layer and the velocity shear interface between the two layers is much more intensive for very dense distributed roughness elements and the two-dimensional turbulence which would be generated at the interface shear layer due to the superimposed barotropic flow decays much more faster in three-dimensional turbulence in such cases. These results are summarized also in table 6.2.

This also has shown to have ripercussions on the different behaviors relative to the entrainment coefficients. Reduced entrainment coefficients have been reported only for the experiment with a less density of roughness elements. In the smooth case, the main source of entrainment is given by the wake generated in the lee of the surge-like flows induced by the barotropic component. In the dense roughness case, with high distributed roughness elements, the main contribution to entrainment is given by the bottom roughness, which assures a very intensive interaction between the bottom boundary layer and the interface between the two layers. In the third experiment, with a low density of roughness elements, the major contribution to the entrainment comes from the bottom roughness: but a reduced interaction between the boundary layer and the interface and the hindered collapsing phase of the interfacial large-scale structures generated by the pulsating barotropic flow both leave to the reduction of the entrainment coefficients. These results are summarized also in table 6.2.

The study of the local effect of roughness in stratified exchange flows is important as it well represents the natural conditions in environmental ambient as in ocean straits or in the atmosphere between valleys, influencing the exchanged masses of water and the air circulation.

7 Linear stability of accelerating two-layer shear flows

7.1 Introduction

Linear stability analysis is a useful tool for predicting the type of instability and its occurence under different flow conditions. The traditional method adds a perturbation to a background flow and the governing equations are then linearized about this state. The normal mode approach allows one to determine the wave number, the growth rate, and the spatial structure of the most unstable modes. Herein it is intrinsically assumed that if the initial perturbation from the mean flow is white noise, then it is the most unstable mode that grows fastest and dominate.

When linear stability analysis was first used, many investigators hoped that the transition to turbulence would follow directly from linear stability analysis. It was quickly shown, however, that this is not necessarily the case. Linear stability analysis is only valid for a very short time before nonlinear effects become important. Nevertheless, linear instability does correctly describe the onset and early evolution of infinitesimal perturbations. It also seems to give a qualitatively correct indication of the overall stability of the flow [Maslowe, 1985]. For this reason much of the literature has been devoted to linear stability theory [Gage, 1973, Pellacani, 1976, Drazin and Reid, 1981].

The normal mode approach of linear stability analysis reduces the system of nonlinear partial differential equations (PDE) into a system of linear ordinary differential equations (ODE) where the boundary conditions depends on the flow configuration. These new differential equations describe an eigen-problem with eigenvalues. However, this transformation from PDE into ODE does not come without complications. Namely, with this substitution, the complex eigenfunctions are unknown functions of the spatial variable and it introduces unknown eigenvalues. Hence, one must make assumptions concerning these unknowns in order to obtain any solution. In one case, one may assume that the disturbance amplifies in space and not in time with a fixed frequency. As such α becomes the unknown eigenvalue, $\omega_i = 0$ and ω_r is specified. This is referred to as *spatial stability theory*. A second case could assume that the disturbances amplify in time and not in space. Hence, ω becomes the unknown eigenvalue $\alpha_i = 0$ and α_r is specified. This is referred to as *temporal stability theory*. The intersection of temporal and spatial theories occur at neutral locations where disturbances neither amplify nor decay in space *and* time. That is $\alpha_i = \omega_i = 0$ and both theories yield the same normal mode solution.

In the spatial stability theory with α complex and ω real, it should be noted that ω can be real if and only if the phase speed $c = \omega/\alpha$ is also complex. Thus, we have that $\omega_r = \alpha_r c_r - \alpha_i c_i$ and $\omega_i = \alpha_r c_i + \alpha_i c_r = 0$ which is much more complicated from a mathematical point of view and this becomes even more significant when viscous effects are considered. This is the reason because most of the stability theory has been devoted to study the stability from a temporal frame of reference and then relating the temporal results to spatial modes by means of the phase velocity, i.e. $\omega_i = -c_r\alpha_i$. Such a simple relation is not valid in general. Gaster corrected the relationship by making use of the group velocity instead of the phase velocity. Gaster [1962] noted that there exist an asymptotic relation between temporally increasing and spatially increasing disturbances. After Gaster [1962], if the temporal solution has parameters $\alpha = \alpha_r$ and $\alpha c = \alpha_r c_r + \imath\alpha_r c_i$, then the corresponding spatial solution has parameters $\alpha = \alpha_r - \imath\alpha_r c_i/c_g$ and $\alpha c = \alpha_r c_r$, where $c_g = \partial(\alpha_r c_r)/\partial\alpha_r$ is the group velocity. This relation is now referred to as Gáster transformation in honor of this important contribution to hydrodynamic stability theory. Temporal studies can be so qualitatively extended to spatially developing flows and can be extrapolated with the above relations in cases where the convective velocity is large in comparison to the mean shear.

However, it is well known that spatial stability theory is more appropriate to describe field or laboratory situations, as a reference frame is singled out by boundary conditions and therefore one should refer to the concept of absolute or convective instability to understand the dynamics of the flow (see Huerre and Monkewitz [1990]). Convectively unstable flows are known to behave as noise amplifiers and their dynamics are described by the spatial stability theory. In contrast, absolutely unstable flows exhibit self sustained oscillations even in the absence of external perturbations, since the zero group velocity wave is amplified in the selected frame.

The following sections are organized as follows: firstly, the governing stability equation will be derived for a two-layer, stratified flow accelerating down a slope (section 7.2). Herein it is assumed that the flow is parallel, incompressible and the Boussinesq approximation is applied. The employed velocity and density profiles will be defined in section 7.3, with the relevant scales for the non-dimensionalization. A brief summary on the existing most important results will be summarized in section 7.4. Finally the employed numerical solution methods (section 7.5) and the results on the linear stability analysis (section 7.6) performed will be presented.

7.2 Governing equations

A two-dimensional, unsteady, stratified flow can be described completely using the velocity components in the horizontal x and vertical z directions, u and w respectively, the density ρ, the concentration of the stratified agent S and the pressure p. For a Newtonian, incompressible fluid flowing over an inclined wall (θ) and assuming a linear equation of state, the governing equations are given by (cf. figure 7.1):

- Conservation of momentum:

$$\rho(u_t + uu_x + wu_z) = -p_x + \rho g sin\theta + \mu(u_{xx} + u_{zz}), \tag{7.1a}$$

$$\rho(w_t + uw_x + ww_z) = -p_z - \rho g cos\theta + \mu(w_{xx} + w_{zz}), \tag{7.1b}$$

- Conservation of mass:

$$\rho_t + (\rho u)_x + (\rho w)_z = 0, \tag{7.2}$$

- Diffusion equation for stratifying agent:

$$S_t + uS_x + wS_z = \kappa(S_{xx} + S_{zz}), \tag{7.3}$$

- Equation of state

$$\rho = \rho_0[1 - \gamma(S - S_0)], \tag{7.4}$$

where g, μ, κ, and γ are the gravitational acceleration, the molecular viscosity, the diffusivity of the stratifying agent and the coefficient of expansion for the stratifying agent respectively, and they are all assumed to be constant. The subscript 0 represents the variables in a standard state and the subscripts x or z denotes differentiation.

For an incompressible flow the change of density due to transport of stratifying agent does not affect the mass conservation, so that equation (7.2) leaves to the simple form of the continuity equation:

$$u_x + w_z = 0. \tag{7.5}$$

Equation (7.3) and equation (7.4) can be combined to give

$$\rho_t + u\rho_x + w\rho_z = \kappa(\rho_{xx} + \rho_{zz}). \tag{7.6}$$

If ρ and p are expanded about a state of hydrostatic equilibrium ρ_0 and p_0, the equations of conservation of momentum can be written as

$$\rho(u_t + uu_x + wu_z) = -p'_x + \rho' g sin\theta + \mu(u_{xx} + u_{zz}), \tag{7.7a}$$
$$\rho(w_t + uw_x + ww_z) = -p'_z - \rho' g cos\theta + \mu(w_{xx} + w_{zz}), \tag{7.7b}$$

where $\rho = \rho_0 + \rho'$, $p = p_0 + p'$, $dp_0/dz = -g\rho_0 \cos\theta$, and where the prime denotes dynamic quantities. If the density difference is small compared to the standard density ρ_0, the Boussinesq approximation can be used and the momentum equations can be rewritten as

$$u_t + uu_x + wu_z = -\frac{p'_x}{\rho_0} + \frac{\rho'}{\rho_0} g sin\theta + \nu(u_{xx} + u_{zz}), \tag{7.8a}$$
$$w_t + uw_x + ww_z = -\frac{p'_z}{\rho_0} - \frac{\rho'}{\rho_0} g cos\theta + \nu(w_{xx} + w_{zz}), \tag{7.8b}$$

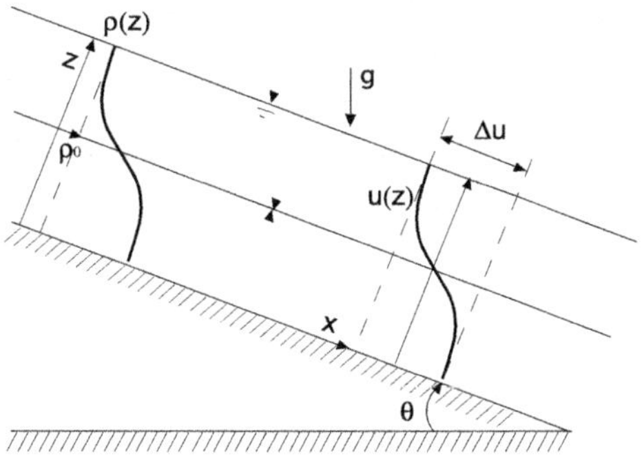

Figure 7.1: Definition for deriving the governing stability equation.

where $\nu = \mu/\rho_0$ is the kinematic viscosity. These equations together with the continuity equation and equation (7.6), are valid for small density variations and for incompressible flows.

The continuity equation for incompressible flows can be used to define a stream function ψ such that

$$u = \frac{\partial \psi}{\partial z} \quad ; \quad w = -\frac{\partial \psi}{\partial x} \tag{7.9}$$

Differentiating the momentum equation in x direction with respect to z and the momentum equation in z direction with respect to x and subtracting the second resultant equation from the first, one obtains the following equation for the stream function.

$$(\triangle\psi)_t + u(\triangle\psi)_x + w(\triangle\psi)_z = \frac{g}{\rho_0}(\rho'_x \cos\theta + \rho'_z \sin\theta) + \nu\triangle\triangle\psi \tag{7.10}$$

where $\triangle = \partial^2/\partial x^2 + \partial^2/\partial z^2$ is the two-dimensional Laplace operator. Non-dimensionalization of equation (7.10) is done by introducing the following non-dimensional dimensional variables:

$$\begin{array}{lll} u = Uu^* & \psi = UL\psi^* & x = Lx^* \\ w = Uw^* & \rho = \varrho\rho^* & z = Lz^* \\ & & t = L/Ut^* \end{array}$$

with U, L, ϱ being a velocity scale, a length scale and a density scale, respectively. These scales will be defined later. Substituting the above definitions into the governing equations (7.10) and (7.6), one obtains

$$(\triangle\psi)_t + u(\triangle\psi)_x + w(\triangle\psi)_z = J(\rho'_x \cos\theta + \rho'_z \sin\theta) + \frac{1}{Re}\triangle\triangle\psi \tag{7.11a}$$

$$\rho_t + u\rho_x + w\rho_z = \frac{1}{PrRe}\triangle\rho. \tag{7.11b}$$

where the $*$ notation has been dropped for simplicity. From the above equations three non-dimensional parameters can be recognized: the bulk (local) Richardson number J, the Reynolds number Re and the Prandtl number Pr. These parameters are defined as

$$J = \frac{\varrho g L}{\rho_0 U^2}, \qquad Re = \frac{UL}{\nu}, \qquad Pr = \frac{\nu}{\kappa}. \tag{7.12}$$

where the Prandtl number gives the ratio between the kinematic viscosity and the diffusion coefficient of the stratified agent and is also known as the inverse of the Schmidt number Sc.

For the linear stability analysis, the flow field is splitted into a parallel mean component and a perturbation field, i.e.

$$\psi'(x,z,t) = \hat{\psi}(z) + \tilde{\psi}(x,z,t), \tag{7.13a}$$
$$\rho'(x,z,t) = \hat{\rho}(z) + \tilde{\rho}(x,z,t), \tag{7.13b}$$
$$u'(x,z,t) = \hat{u}(z) + \tilde{u}(x,z,t), \tag{7.13c}$$

Substituting these expressions into the governing equations (7.11), and collecting terms of equal magnitude (neglecting higher order terms) and subtracting the background flow, the perturbation equations can be obtained:

$$(\Delta\tilde{\psi})_t + \hat{u}(\Delta\tilde{\psi})_x + w'\hat{u}_{zz} = J(\tilde{\rho}_x \cos\theta + \tilde{\rho}_z \sin\theta) + \frac{1}{Re}\Delta\Delta\tilde{\psi}, \tag{7.14a}$$

$$\tilde{\rho}_t + \hat{u}\tilde{\rho}_x + w'\hat{\rho}_z = \frac{1}{PrRe}\Delta\tilde{\rho}. \tag{7.14b}$$

Using the method of normal modes

$$\tilde{\psi}(x,z,t) = \Re[\phi(z)e^{\imath\alpha(x-ct)}], \tag{7.15}$$

$$\tilde{\rho}(x,z,t) = \Re[\varphi(z)e^{\imath\alpha(x-ct)}] \tag{7.16}$$

and substituting these expressions into the linear equations (7.14a) the eigen-problem can be obtained

$$[(\hat{u}-c)(\phi_{zz}-\alpha^2\phi)] - \hat{u}_{zz}\phi = J\left[\varphi\cos\theta + \frac{\varphi_z}{\imath\alpha}\sin\theta\right] + \frac{1}{Re\imath\alpha}[\phi_{zzzz} - 2\alpha^2\phi_{zz} + \alpha^4\phi]. \tag{7.17}$$

Similarly, equation (7.14b) gives

$$(\hat{u}-c)\varphi - \hat{\rho}_z\phi = -\frac{1}{\imath\alpha PrRe}(\varphi_{zz} - \alpha^2\varphi + \hat{\rho}_{zz}). \tag{7.18}$$

Herein, ϕ denotes the complex amplitude of the disturbance, $\alpha = \alpha_r + \imath\alpha_\imath$ and $\omega = \omega_r + \imath\omega_\imath$ are the wave number, the spatial amplification rate, the frequency and the temporal amplification rate of the perturbation, respectively and $c = \omega/\alpha = c_r + \imath c_i$ is the complex wave speed. Equation (7.17) and equation (7.18) can be solved to study the linear stability of the flow.

Note that without stratification ($J = 0$), equation (7.17) reduces to the Orr-Sommerfeld equation.

If the density diffusion is neglected ($Pr \to \infty$), the right hand side of equation (7.18) can be neglected and

$$\varphi = \frac{1}{\hat{u}-c}\hat{\rho}_z\phi \tag{7.19}$$

Introducing this expression into equation (7.17) one obtains:

$$(\phi_{zz} - \alpha^2\phi) - \frac{u_{zz}}{(u-c)}\phi + J\cos\theta\frac{\rho_z}{(u-c)}\phi + \frac{J\sin\theta}{\imath\alpha(u-c)^2}\left[\rho_{zz}\phi - \frac{\rho_z u_z}{(u-c)}\phi + \rho_z\phi_z\right] = \frac{1}{\imath\alpha Re(u-c)}[\phi_{zzzz} - 2\alpha^2\phi_{zz} + \alpha^4\phi] \tag{7.20}$$

where the $\hat{}$ has been dropped for simplicity. This is the final form of the governing equation, valid for a stratified, incompressible, parallel, Boussinesq, shear flow. Herein, $u(z)$ and $\rho(z)$ are the velocity and density profiles, respectively, which need to be defined a priori (see section 7.3). Note that without slope ($\theta = 0$) and neglecting viscous effects, equation (7.20) reduces to the Taylor-Goldstein equation. J, Re, α and c are the unknowns (four) variables. By fixing two of them, the other two can be numerically determined.

Finally, the definition of the different velocity, density and length scales need to be defined. To be consistent with the definitions of the non-dimensional velocity and density profiles defined in section 7.3, we choose the following scales:

$$U \equiv \frac{\Delta U}{2}, \qquad \varrho \equiv \frac{\Delta\rho}{2}, \qquad L \equiv \delta_\nu, \tag{7.21}$$

where $\Delta U = U_2 - U_1$ is the velocity difference between the two layers, $\Delta\rho = \rho_2 - \rho_1 = -(\rho_1 - \rho_2)$ (this explains the change of sign of the terms containing the bulk Richardson number J) is the density difference between the two layers, and δ_ν is the thickness of the velocity shear layer (see figure 7.2). With these definitions, the three adimensional parameters reads

$$J = \frac{2\Delta\rho g\delta_\nu}{\rho_0(\Delta U)^2}, \qquad Re = \frac{\Delta U\delta_\nu}{2\nu}, \qquad Pr = \frac{\nu}{\kappa}. \tag{7.22}$$

In the remainder of this study, we will deal with the inviscid stability equation and the term containing the Reynolds number will be neglected.

7.3 The velocity and density profiles

The velocity and density profiles are usually considered in a dimensionless form, but there are slight differences in the ways different authors achieve the non-dimensionalization. This can lead to misunderstandings in effecting comparisons.

Here we take first the dimensional velocity and density profiles:

$$u(z) = A + \frac{\Delta U}{2} tanh\left(\frac{z}{\delta_\nu}\right), \tag{7.23}$$

$$\rho(z) = \rho_0 + \frac{\Delta\rho}{2} tanh\left(\frac{z}{\delta_\rho}\right), \tag{7.24}$$

where A is a superposed background flow, $\Delta U = U_2 - U_1$ is the velocity difference between the two layers, $\rho_0 = (\rho_1 + \rho_2)/2$ is a reference density. This contribute can be dropped as it is a constant and only the first and second derivatives of the density profiles are required in the stability governing equation (7.20). The typical vertical scales must not be necessarily the same for the velocity and density profiles: thus, two different typical vertical scales are defined, namely δ_ν for the velocity profile and δ_ρ for the density profile. The profiles are made dimensionless by use of the velocity scale $\Delta U/2$, the density scale $\Delta\rho/2$ and as vertical length scale δ_ν. The profiles reads then

$$u^*(z) = a + tanh(z^*), \tag{7.25}$$

$$\rho^*(z) = tanh(Rz^*), \tag{7.26}$$

where the stars denotes non-dimensional quantities, $a = 2A/\Delta U$ is the non-dimensional background flow and $R = \delta_\nu/\delta_\rho$ is the ratio of the velocity shear layer thickness to the density layer thickness and changes the steepness of the density profile.

Hazel [1972] proposed after definition of the typical vertical length scale δ_ν, the following definitions for the velocity and density profiles:

$$u(z) = Vu^*(z^*) \qquad \text{such that} \qquad (du^*/dz^*)_{z^*=0} = 1, \tag{7.27}$$
$$log(\rho/\rho_0) = \lambda(z^*) = \sigma\rho^*(z^*) \qquad \text{such that} \qquad (d\rho^*/dz^*)_{z^*=0} = 1 \tag{7.28}$$

with V being a typical velocity, σ a typical density measure. Remembering the definition of the gradient Richardson number $Ri(z)$ with the above definitions, we have:

$$Ri(z^*) = \frac{Jd\rho^*/dz^*}{(du^*/dz^*)^2} \tag{7.29}$$

where $J = \sigma g\delta_\nu/V^2$ is the bulk Richardson number. The use by Hazel [1972] of this particular scaling for the velocity and density profiles, ensures that the dimensionless number J is always equal to the overall Richardson number at the origin. Hazel [1972] studied the stability as a function of R defining then the velocity and density profiles as follows:

$$u^*(z^*) = tanh(z^*) \tag{7.30}$$
$$\rho^*(z^*) = \frac{1}{R}tanh(Rz^*). \tag{7.31}$$

These definitions for the velocity and the density profiles assure the conditions given in (7.28) and thus, that the gradient Richardson number Ri and the Richardson number at the origin J are equal. This definition is in some cases unphysical because the density ranges between $\pm 1/R$ and not between the constant value ± 1, but has the merit of uniformity for comparisons between different cases. In the following, the stars indicating non-dimensional quantities will be dropped for simplicity.

With the definitions of our profiles given in equation (7.26), the condition that the first derivative of the non-dimensional density profile is not satisfied, so that the equality between the overall and local Richardson numbers does not hold.

The derivatives of the velocity and density profiles (7.26) are given by

$$u_z = 1 - \tanh^2(z) \qquad u_{zz} = -2\tanh(z)u_z \tag{7.32}$$
$$\rho_z = R(1 - \tanh^2[Rz]) \qquad \rho_{zz} = -2R\rho\rho_z \tag{7.33}$$

and are plotted in figure 7.2 (b).

Pawlak and Armi [1998] also used the same definitions for the density and velocity profiles given by Hazel [1972] in order to satisfy the conditions given in (7.28), with only one difference in the definition of the velocity profile, where a background flow has been added.

Most of the other authors (Lawrence and Redekopp [1991], Haigh and Lawrence [1999], Ortiz and Loiseleux [2002] with a background flow) used the piecewise linear velocity and density profiles, without satisfying the rule (7.28) proposed by Hazel [1972].

Here, the choice of the hyperbolic tangent profiles for both the velocity and density distributions is justified by the good agreement between the theoretical definitions and the experimental data, as demonstrated in figure 7.2 (c) and (d).

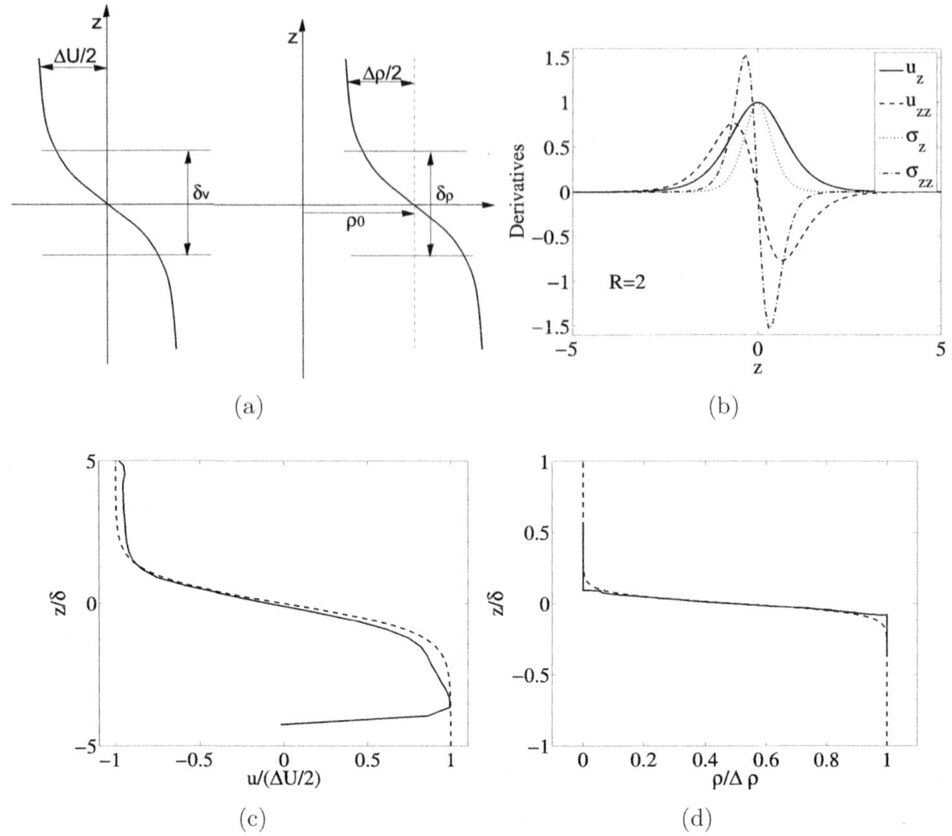

Figure 7.2: Sketch of typical velocity and density profiles for a stratified exchange flow taken from (Zhu and Lawrence 2001) (a) and first and second derivatives of the velocity and density profiles (for $R = 2, \epsilon = 0.5$)(b). Comparison between the experimental (continuous line) and theoretical (dashed line) velocity (c) and density (d) profiles.

7.4 A brief summary of previous stability studies on stratified shear flows

7.4.1 Stability studies from a temporal frame of reference

Miles [1961] have proven that stability of an inviscid continuous stratified flow is assured if the gradient Richardson number, which gives the ratio between buoyancy to intertia forces, is everywhere greater that 1/4. However, Maslowe [1985] have shown that stratification effects are in general more complex since stable stratification adds a restoring force that constrains the vertical displacement of particles, and the stability depends on the details of the density and velocity profiles. A large effort has been devoted to understanding the effect of buoyancy forces on shear instability by experimental and theoretical studies (Holmboe [1962], Hazel [1972], Koop and Browand [1976], Smyth and Peltier [1988, 1989, 1991], Lawrence and Redekopp [1991], Haigh and Lawrence [1999], Strang and Fernando [2001], Zhu and Lawrence [2001], Hogg and Ivey [2003]). Based upon the totality of

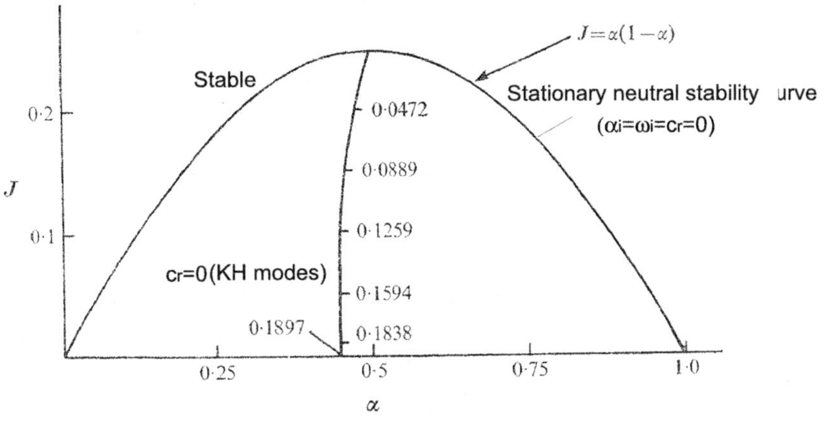

Figure 7.3: Stationary neutral stability curve for equal shear and density shear layer thicknesses ($R = 1$). The stationary stability curve is also the stability boundary. Inside the curve, KH modes develop characterized with a zero phase speed. From Hazel [1972])

these studies, which have focused on temporal stability of the stratified shear flows, it is known that if the characteristic thicknesses of the velocity shear and density interface are similar, the stability is **stationary** with respect to the mean flow (i.e. the phase velocity $c_r = 0$) and is called Kelvin-Helmholtz (KH) instability. For a piecewise velocity profile and a two layer density step, Holmboe [1962] found theoretically that the shear layer is primarily unstable either to KH waves *or* to two travelling waves, which have the same amplification rates but opposite propagating directions with respect to the mean speed of the shear layer. These waves are known as Holmboe waves. In contrast to KH modes, Holmboe waves are not restabilized when the stratification increases in the inviscid approximation. When the bulk Richardson number (the value of the Richardson number at the origin of the vertical coordinate system) is low enough, KH modes will develop, while if the bulk Richardson number is increased, the development of KH modes is inhibited and the interface starts being deformed by Holmboe waves. Assuming tangent hyperbolic profiles, Hazel [1972] studied the stability as a function of the ratio of the scale of the shear thickness to the density interface thickness. He showed applying the Miles-Howard (Miles [1961]) criterion that a ratio greater than two is a necessary condition for instability whatever the value of the bulk Richardson number. Generally, the transition between stable and unstable modes is given by the stationary neutral stability curve, where $\alpha_i = \omega_i = 0$. This holds also for the stability of stratified shear flows having the same thickesses of the velocity and density shear layers (cf. figure 7.3). However, if the thicknesses of the shear and density layer are not the same, i.e. for ratios equal or larger than two, the stationary neutral stability curve where $\alpha_i = \omega_i = 0$ and where the phase speed $c_r = 0$ is **not** the stability boundary, but represents the transition from KH modes to Holmboe modes, characterized with a phase speed $c_r \neq 0$ (Hazel [1972]). Referring to figure 7.4 in which different regions are defined with letters, the results are summarized as in Hazel [1972] for different thicknesses of the velocity and density shear layers, as follows:

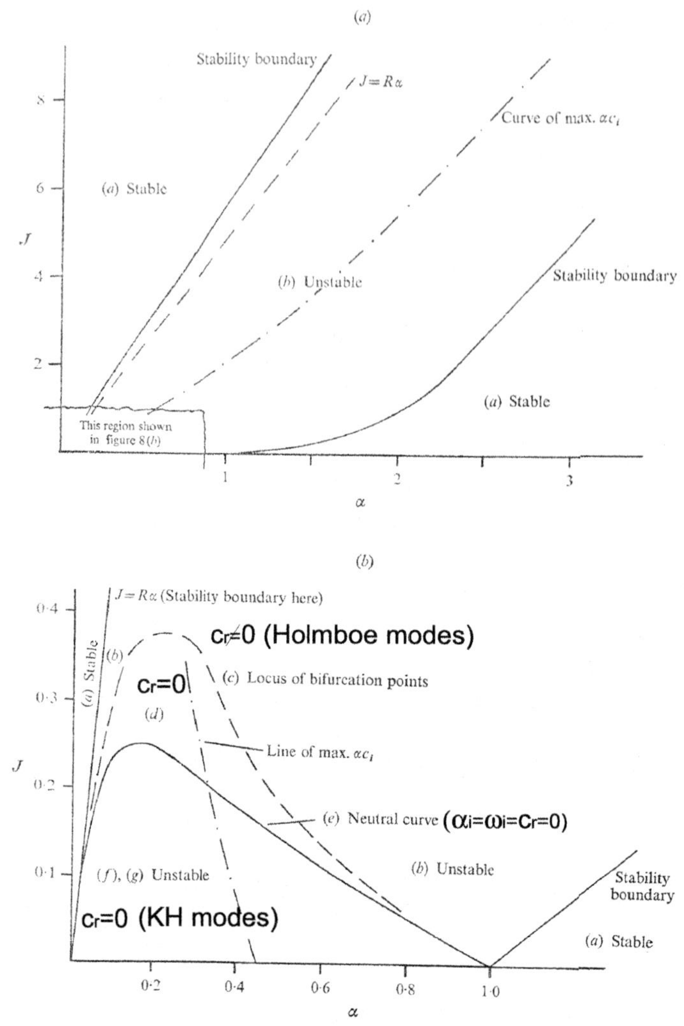

Figure 7.4: Stability characteristics for different shear and density shear layer thicknesses ($R = 5$). The stationary neutral stability curve is not the stability boundary. Inside the continuous curve, KH modes develop, characterized with a zero phase speed, outside Holmboe modes develop, with a non zero phase speed. From Hazel [1972]).

- In the regions (a) no solutions could be found by Hazel [1972], so that only neutral modes lives in this region.
- In region (b), an unstable moving mode has been found ($c_r \neq 0, c_i > 0$). Because of the anti-symmetry of the velocity and density profiles due to different values of the thicknesses (i.e. in figure 7.4, $R = 5$) there are in fact two modes, of the same growth rate, moving with equal velocity in opposite direction (Holmboe modes).
- In region (d), there are two stationary, unstable modes ($c_r = 0, c_i > 0$, KH modes).
- The dividing line (c), between regions (b) and (d), marks the locus of bifurcation

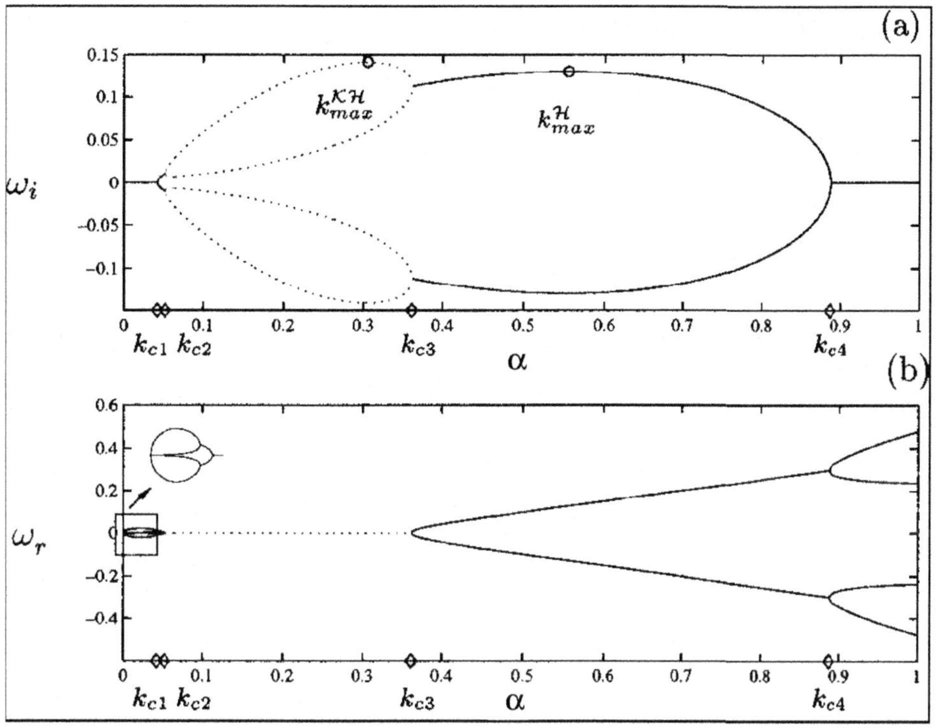

Figure 7.5: Stability characteristics for different shear and density shear layer thicknesses ($R = 5$). The stationary neutral stability curve is not the stability boundary. Inside the continuous curve, KH modes develop, characterized with a zero phase speed, outside Holmboe modes develop, with a non zero phase speed. From Ortiz and Loiseleux [2002]).

points in the c_r, c_i-plane, points where the phase velocities of the two Holmboe modes of region (b) become zero, and the growth rates of all four modes of regions (b) and (d) are equal. This is well illustrated in figure 7.5 (from Ortiz and Loiseleux [2002]) where the temporal amplification rates ω_i and the wave frequency ω_r are plotted versus the real wave number α_r. Herein, KH modes are represented in dotted lines and the Holmboe modes in continuous lines, for $Ri = 0.04$ and using piecewise linear velocity and density profiles instead of the tangent hyperbolic profiles.

- In regions (f) and (g) there is one stationary, unstable mode (KH mode).

- The dividing line between the regions (d) and (f) is the stationary neutral curve where $\alpha_i = \omega_i = c_r = 0$. This relates to the other stationary, unstable mode, which is unstable outside it, in region (d).

Other authors investigated the effect of an offset of the density interface relative to the shear interface and found that this results in the formation of asymmetric Holmboe waves. For more details see Haigh and Lawrence [1999], Zhu and Lawrence [2001].

7.4.2 Stability studies from a spatial frame of reference

Only recently has the spatially stability theory been addressed by Pawlak and Armi [1998] in the case of wedge flow where the upper stream was not moving and in which the ratio between the velocity shear layer and the density shear layer thicknesses was very large ($R = 50$) for small Richardson numbers. They clearly demonstrated that the spatial theory differs strongly from the temporal theory. The most amplified modes are different and the spatial instability results from a combination of KH modes and Holmboe instabilities, which was called in their study a *hybrid* instability region. Later, Ortiz and Loiseleux [2002] extended the analysis by Pawlak and Armi [1998], by analyzing the effect of the mean advection and of changing the Richardson number, for piecewise velocity and density profiles. Moreover they first studied the transition between the absolute and convective instability, showing that the classical scenario for the transition should be modified due to the presence of propagating waves (cf. figure 7.6). In the convective region, the spatial theory is relevant and the slowest propagating wave is shown to be the most spatially amplified. These results of their study can be summarized as follows (see figure 7.7): when both streams move in the same direction ($a > 1$), the flow is convectively unstable whatever the value of the Richardson number. When the streams propagate in opposite direction ($a < 1$) two cases must be distinguished: when both the KH and Holmboe are unstable under a certain value of the Richardson number, the flow is always absolute unstable. When only Holmboe waves are unstable, over the Richardson number limit, both convective and absolutely instabilities can live. This means that in an equal exchange flow, like that treated in this study, when the Richardson number is lower than a certain threshold the flow will be only absolutely unstable, while for a Richardson number larger than this threshold, the flow will be only convectively unstable. However, as pointed out in Ortiz and Loiseleux [2002], absolutely instabilities could arise due to self-sustained resonance triggered by reflective boundary conditions.

Finally, they showed that the spatial and temporal results differ from each other remarcably only for values of the normalized background flow greater than $a \sim 0.2$.

More recently, Gelfgat and Kit [2006] studied the spatial instability of parametrically excited stratified mixing layer flows together with the related temporal instability problem. They performed a parametric analysis of the temporal and spatial Kelvin–Helmholtz and Holmboe instabilities and compared characteristic features of the instabilities, including results for the viscid solution. Moreover, they showed that in their study, the Gaster transformation is valid for the Kelvin–Helmholtz instability, but cannot be applied to the Holmboe one. Herein, it was also found that for the same governing parameters the spatial upstream and downstream Holmboe waves have different amplification rates and different absolute phase velocities, with larger difference observed at larger Richardson numbers.

In an exchange flow like that treated in this study, $a = 0$, and the temporal and spatial results differ very weakly. Thus, we will address the stability analysis from a **temporal** frame of reference.

No linear stability studies studying the influence of a decomposition of the buoyancy term in the governing stability equation in one direction parallel to an inclined boundary and in one direction orthogonal to this inclined wall, are available.

Goal of the linear stability analysis in this work, is to examine the influence of the spa-

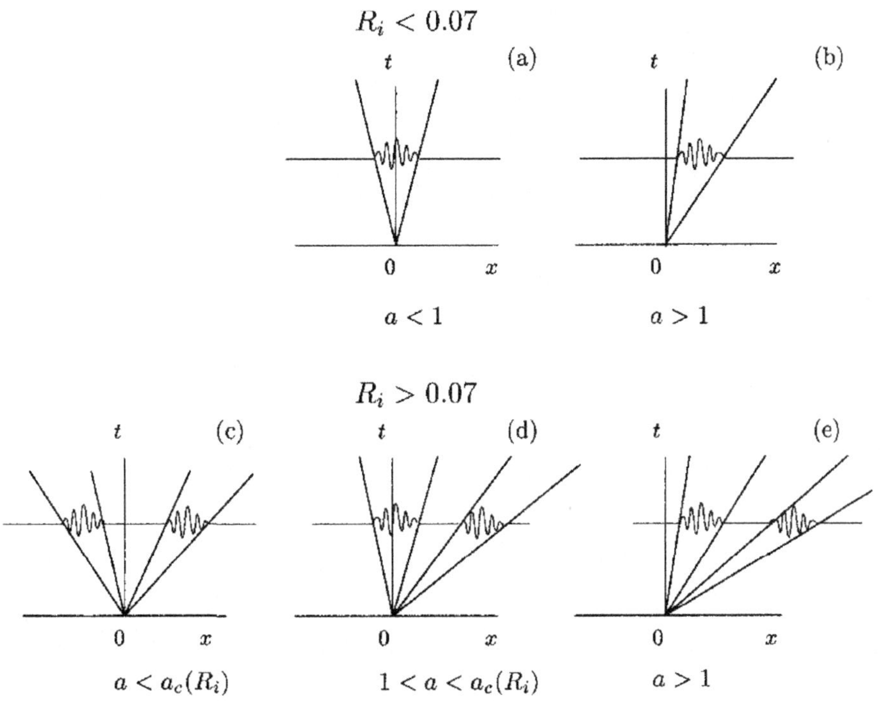

Figure 7.6: Sketch of the impulse responses in the (x, t)-plane. For $Ri < 0.07$ a single wave packet, (a) absolutely unstable for a normalized background flow $a < 1$ and (b) convectively unstable for $a > 1$. For $Ri > 0.07$ behavior of the two Holmboe wave packets for a positive varying as follows: (c) convectively unstable, (d) absolutely unstable and (e) convectively unstable. From Ortiz and Loiseleux [2002]).

tial acceleration (sill slope) on the stability of the interface under different flow conditions, i.e. under the influence of a weak superposed background flow a.

7.5 Numerical solution methods

As the coefficients of the resulting ODE, i.e. the governing stability equation, are in general not constant, they can be solved exactly for only a handful of special cases. These are mostly limited to steady flows approximated by either piecewise constant or linear velocity and density profiles (see Drazin and Reid [1981], Criminale and Joslin [2000]). Although these flows are not physically realistic, they often admit simple solutions. For more complicated flow profiles it is usually necessary to use either asymptotic approximations, or to solve the equations numerically.

We propose in the following two different numerical solution methods to solve the eigenvalue problem: the first is based on a power series expansion of the amplitude function ϕ in Chebyshev Polynomials, while the second method is based on a centered finite difference method.

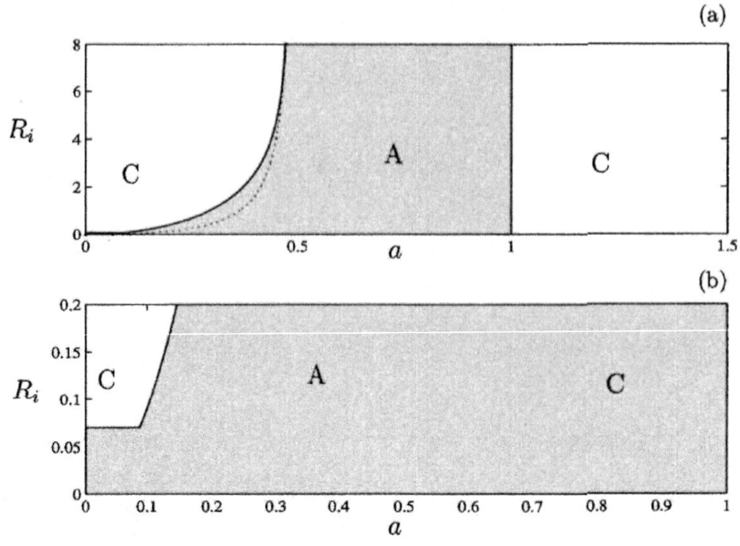

Figure 7.7: (a) in light gray absolutely unstable modes (A) in white convectively unstable modes (C) for different Richardson numbers and versus different values of the normalized background flow a. From Ortiz and Loiseleux [2002]).

7.5.1 Pseudospectral method

A pseudo-spectral collocation method can be used to solve equation 7.20, which employs Chebyshev Polynomials, mapped to an infinite domain, which are more appropriate for a shear interface, since no truncation is required to apply the boundary conditions at $\pm\infty$ ([Canuto et al., 1988]).

Here, Chebyshev polynomials are used as basis functions for the expansions of ϕ. To use Chebyshev polynomials, the transverse direction is mapped into the interval $[-1, 1]$. It is assumed that as the space variable goes to infinity, ϕ approaches some known continuous function of the other space variables and time. In other words, the boundary conditions for ϕ at infinity are known.

Here, the definition of the Chebyshev polynomials is given by

$$T_k(z) = \cos(k \cos^{-1} z), \tag{7.34}$$

which has collocation points

$$z_j = -\cos(\pi(2j-1)/(2(N-2)), \tag{7.35}$$

where N denotes the number of grid points. With this definition, the Chebyshev polynomials are defined on $[-1, 1]$. The highest density of collocation points, formed by this function, is located at the boundaries. Boyd [1989] suggested the following mapping function

$$\xi_j = \frac{\gamma z_j}{\sqrt{1 - z_j^2}}, \tag{7.36}$$

defined on $[-\infty, \infty]$, to map z_j to an unbounded domain and with γ defining the density of grid points over the domain. So, the highest density of collocation points is found at the origin. This is an advantage for a shear interface since the strongest velocity gradients occur where the grid is the most refined.

To use the mapped Chebyshev polynomials, $TB_n = T_n(\xi)$, an equation in the following form is needed

$$\chi_1(z)\phi'''' + \chi_2(z)\phi'' + \chi_3(z)\phi' + \chi_4(z)\phi = 0. \tag{7.37}$$

Then, an approximation of the eigenfunction is sought as

$$\phi(z) = \sum_{k=0}^{N} \hat{\phi}_k TB_k(z), \tag{7.38}$$

where $\hat{\phi}_n$ are constant coefficients. The collocation equations are, then,

$$\sum_{k=0}^{N} \hat{\phi}_k \left[\chi_1(z_k)TB_k'''' + \chi_2(z_k)TB_k'' + \chi_3(z_k)TB_k' + \chi_4(z_k)TB_k\right] = 0, \tag{7.39}$$

with the boundary conditions

$$\phi(\pm\infty) = \phi'(\pm\infty) = 0. \tag{7.40}$$

By the definition of the Chebyshev polynomials, they are satisfied automatically, so that they must not be imposed in the numerical solution. Rearranging the terms in equation (7.20), the coefficients χ_i $(i = 1, 2, 3, 4)$ in equation (7.37) are evaluated as

$$\chi_1 = \frac{1}{\alpha \imath Re(u-c)}, \tag{7.41}$$

$$\chi_2 = 1 + \frac{2\alpha^2}{\imath Re(u-c)}, \tag{7.42}$$

$$\chi_3 = +\frac{\rho_z}{(u-c)^2}\frac{J\sin\theta}{\imath\alpha}, \tag{7.43}$$

$$\chi_4 = -\alpha^2 - \frac{u_{zz}}{(u-c)} + \frac{J\sin\theta}{\imath\alpha}\left[\frac{\rho_{zz}}{(u-c)^2} - \frac{\rho_z u_z}{(u-c)^3}\right] + \frac{J\cos\theta\rho_z}{(u-c)^2} - \frac{\alpha^3}{\imath Re(u-c)}. \tag{7.44}$$

The collocation equation (7.39) can be written as

$$[\mathcal{L}] \cdot \hat{\Phi} = 0, \tag{7.45}$$

where $\hat{\Phi} = (\hat{\Phi}_0, \hat{\Phi}_1, ..., \hat{\Phi}_N)^T$ and $[\mathcal{L}]$ is the matrix which assure that equation (7.45) satisfies equation (7.39) and which is defined as follows:

$$\mathcal{L}_{i,j} = \chi_1(z_i)TB_{j+1}''''(z_i) + \chi_2(z_i)TB_{j+1}''(z_i) + \chi_3(z_i)TB_{j+1}'(z_i) + \chi_4(z_i)TB_{j+1}(z_i), \tag{7.46}$$

where the prime denotes differentiation with respect to the spatial variable. Equation (7.45) represents an eigenvalue problem which permits, when solved, to determine a pair of eigenvalues, imposing that the determinant of the matrix $[\mathcal{L}]$ matrix is zero.

Gelfgat and Kit [2006] showed that in the case of tangent hyperbolic profiles, the convergence of these series is particularly slow, even if one uses a high number of basis functions for the approximation of the amplitude function. For this reason, they preferred to use a central, finite difference method with a number of nodes varying between 500 and 2,000.

7.5.2 Finite difference method

The basic idea of the finite difference method is to approximate the partial derivatives appearing in the stability equation (7.20) by discrete difference operators. The most common forms of the finite difference with discrete difference operators are forward, backward and central differences, all of which stem from the Taylor´s series:

$$\frac{d\phi(z)}{dz} \approx \frac{\phi_{i+1} - \phi_i}{\Delta z} \qquad \text{forward} \tag{7.47}$$

$$\frac{d\phi(z)}{dz} \approx \frac{\phi_i - \phi_{i-1}}{\Delta z} \qquad \text{backward} \tag{7.48}$$

$$\frac{d\phi(z)}{dz} \approx \frac{\phi_{i+1} - \phi_{i-1}}{2\Delta z} \qquad \text{central} \tag{7.49}$$

(7.50)

where ϕ is the function which has to be approximated. The order of a finite difference scheme is defined according to the lowest order of the neglected Taylor series expansion terms and indicates the behavior of the numerical error with respect to the discretization length. The usage of the Taylor series expansion to derive the expressions (7.50) is based on the assumption of a smooth function. However, if solutions are not smooth, the use of high order schemes can lead to oscillations in the solution and hence can be less accurate than low order methods (see Ferziger and Peric [1996]).

The finite difference method employed in this study is a high-order, central, explicit method. In a few special cases (e.g. for explicit approximations on equidistant grids), the optimal weights are known in closed form. Fornberg [1998] presented two short algorithms for finding the optimal weights for derivatives of any order, approximated to any level of accuracy. He also provided the algorithms which calculate the exact weights for both explicit and implicit approximations for both regular and irregular grid spacings. He proposed very simple recursion relations for the coefficients, so that starting from a trivial value, all the required weights follow recursively from these expressions, thus reducing enormously the computation time. For more details about the derivation of these recursion relations for the weights and for the algorithms it is referred to Fornberg [1988, 1998].

In the following section the two methods (the pseudospectral method described in section 7.5.1 and the finite difference method described above) are employed to solve the governing stability equation (7.20). The results will provide information on which method is the most appropriate to solve the Taylor-Goldstein equation.

7.5.3 Comparison between the two numerical solution methods and sensitivity analysis

The results obtained applying the different numerical solutions methods were first compared in order to choose the best one. As mentioned before, Gelfgat and Kit [2006] studied the stability of stratified shear flows using the viscid Taylor-Goldstein equation and the tangent hyperbolic profiles. They showed that in the case of tangent hyperbolic profiles, the convergence of the Chebyshev polynomials series is particularly slow, even if one uses a high number of basis functions for the approximation of the amplitude function. Additionally, they compared the results obtained using this general Galerkin method with the results obtained using a central finite difference method employing stretching of the grid where necessary, and showed that a minimum number of 500 to 2,000 of nodes was required to obtain acceptable results.

A first numerical analysis showed that in general, the Taylor-Goldstein equation presents many singular points for different values of the eigenvalues, so that the numerical solution becomes more difficult. In fact, when analyzing the Taylor Goldstein equation we note that compared to the Orr-Sommerfeld equation we have terms which are divided by $(u-c)^n$, where $n = 2, 3$. This results in multiple symmetric solutions and a different behavior as compared to the unstratified stability equation. Moreover, the Taylor-Goldstein equation has shown to be very sensitive to discontinuities in the meshgrid, so that a very high number of basis functions for the general Galerkin method or of a very high number of grid points for the finite difference method is required to obtain the right solutions. From the results obtained applying the two different numerical solution methods, we can confirm the results obtained by Gelfgat and Kit [2006] that the Chebyshev series are converging very slow, so that no gain of the time required for the calculations using the pseudospectral method could be achieved. Moreover, the results obtained using the Chebyshev series showed to have a higher noise as compared to the finite difference method, so that the search for the *physical* solutions becomes more difficult. Bayliss and Matkowsky [1995] also showed that a roundoff error in computing derivatives using the Chebyshev differentiation matrix occurs.

Figure 7.8 shows the condition numbers given for different values of the real and imaginary part of the calculated eigenvalue, i.e. ω. The condition number represents the ratio of the largest singular value to the smallest: large condition numbers indicate a nearly singular matrix. Figure 7.8 (a) and (b) shows the results obtained solving the inviscid Taylor-Goldstein equation using a infinite Chebyshev grid with 50 and 150 basis functions, respectively, while in (c) and (d) the results were obtained using the finite difference method with, respectively, 500 and 1,500 grid points. From the figure it can be noted that the number of numerical (unphysical) solutions is increasing if the number of basis functions increases using the Chebyshev polynomials (figure 7.8 (a) and (b)) and, on the other hand, the noise is higher employing the Chebyshev polynomials in the interesting region (black regions at the axis origin) as compared to the finite difference method (figure 7.8 (a) and (c) of (b) and (d)). The numerical, unphysical solutions can be recognized in the figure as the light gray zones, which are also moving depending on the number of basis functions/grid points. The physical solutions (black regions) remain at the same position independently from the numerical grid. Moreover, the time required for the calculation of one eigenvalue is very much larger using the pseudospectral

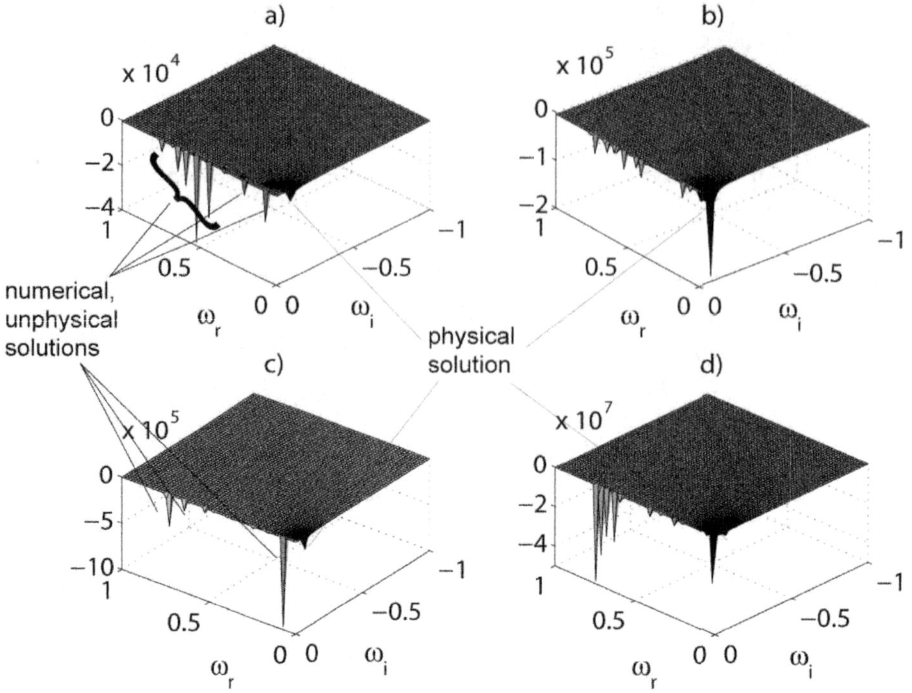

Figure 7.8: Condition numbers for different values of the real and imaginary part of the calculated eigenvalue, i.e. ω. Large condition numbers indicate a nearly singular matrix. Chebyshev polynomials on an infinite grid with (a) 50 basis functions and (b) with 150 basis functions. Central finite difference method with (c) 500 and (b) 1,500 grid points. The noise using the finite difference method is smaller so that the physical solutions are found more easily.

method as compared to the second-order central finite difference method, and also the precision has shown to be lower. In figure 7.9 (a) the value of the calculated eigenvalue (the temporal amplification rate ω_i) for $\alpha_r = 0.5$ and $J = 0.25$ is plotted for different computation times corresponding to a different spatial resolution. The calculated eigenvalue should be zero, as the pair $\alpha_r = 0.5$, $J = 0.25$ lie on the stationary neutral stability curve ($J = \alpha_r(1 - \alpha_r)$, see Holmboe [1962], Hazel [1972]). From the figure it can be clearly seen that the pseudospectral method converges slower and produces less accurate results as compared to the finite difference method on a regular grid spacing, for the same computation times. In figure 7.9 (b), the same temporal amplification rates are plotted versus the number of grid points employed using the finite difference method and the number of basis functions (multiplied by a factor 10) employed using the Chebyshev series, for different values of the wavenumber. The results shows that a minimum number of 10,000 grid points for the finite difference method on a regular grid spacing is required to obtain satisfactorily results.

On the basis of these results from the sensitivity analysis and on the results of Gelfgat and Kit [2006], the second-order finite difference method is chosen. The results shows that a minimum number of 10,000 grid points for the finite difference method is required

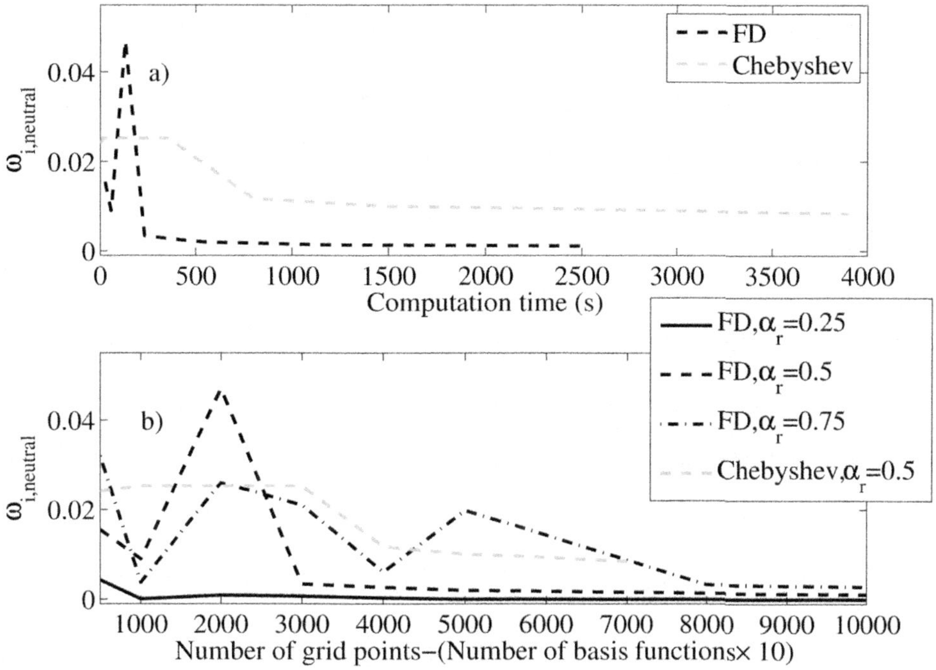

Figure 7.9: Comparison between the results obtained using a second-order central finite difference method on a regular grid spacing and the pseudospectral method employing Chebyshev polynomials. The exact solution for the temporal amplification rates is given by $\omega_i = 0$. a) Temporal amplification rates vs. the computation time using the two numerical solution methods. Chebyshev series show higher computation times and a lower precision. b) Temporal amplification rates versus the number of grid points (finite difference method) or the number of basis functions multiplied by 10 (pseudospectral method) for different values of the wavenumber α_r.

to obtain satisfactorily results, which efforts very high computing times. Due to the high mesh resolution required, an irregular grid will be used with the central finite difference method, which is as the grid defined for the pseudospectral method (infinite grid, see equations (7.35) and (7.36)).

7.6 Results and discussion

7.6.1 Determination of instability types

Hazel [1972] showed that in the unstable region Kelvin-Helmholtz modes develop, characterized by a zero phase speed ($c_r = 0$) and finite temporal amplification rates ω_i, which have shown to decrease for increasing bulk Richardson numbers J. The transition from stable to unstable regions is given by the stationary neutral stability curve, characterized by purely real values of the eigenvalues and additionally, by a zero phase speed. Indeed, Hazel [1972] showed that if $R > 2$, an anomalous behavior starts and the stationary neutral stability curve is not the stability boundary, but the boundary for the transition from KH modes to Holmboe modes, characterized by a non-zero phase speed. This tran-

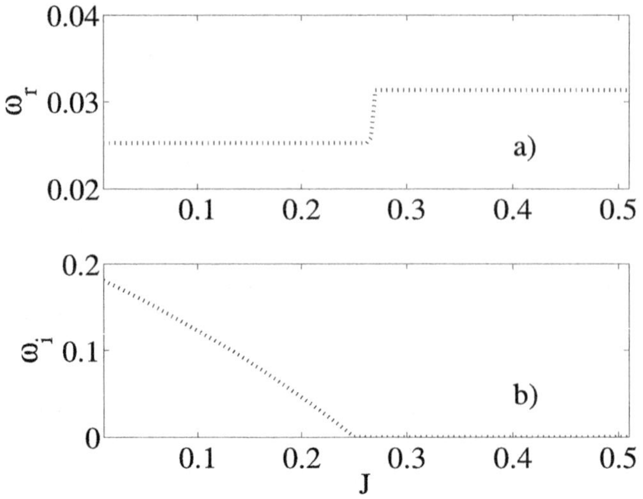

Figure 7.10: Schematic of search criterion for the stability boundaries. a) Wave frequency and b) temporal amplification rates vs. the Richardson number J. The transition is found when the wave frequency changes. At that point, amplification rates changes too.

sition represents the point at which two steady KH modes are replaced by two conjugated oscillatory Holmboe modes and is also known as the Takens-Bogdanov codimension-two bifurcation point (Kuznetsov [1998]). Gelfgat and Kit [2006] showed that this point is also retained for continuous velocity and density profiles and for the complete model that accounts for viscosity and diffusion. Moreover, they also showed that in the spatial case, the KH to Holmboe transition is continuous. In this study, selecting the values for J, R, α_r, and ω_r giving an initial guess for the eigenvalue $\omega = \omega_r + \imath\omega_i$, the numerical code computes the value of ω. We select $R = 1$, a certain wavenumber and let the bulk Richardson number range between $0.01 < J < 0.51$. For every Richardson number, the correspondent eigenvalue ω is computed. This calculation is then repeated for different values of the wavenumber $0.01 < \alpha_r < 1.0$ and for five slope angles, i.e. starting from known results for 0 radians and increasing the slope to reach the slope of 0.2 radians. An example is given in figure 7.10, in which the wave frequency (a) and the temporal amplification rate (b) is plotted versus the bulk Richardson number for no slope angle $\theta = 0$, $R = 1$, $a = 0.05$, and $\alpha = 0.505$. It can be observed that the wave frequency remains at a constant value an then it increases suddenly, reaching a higher constant value. This happens at a Richardson number $J = 0.26$. The amplification rate vanishes at $J = 0.25$. At this point, the transition from unstable modes (KH billows characterized by a zero phase speed) to stable modes occurs. Also, these results were used to validate the chosen numerical method and are in good agreement with the well known results obtained from Hazel [1972].

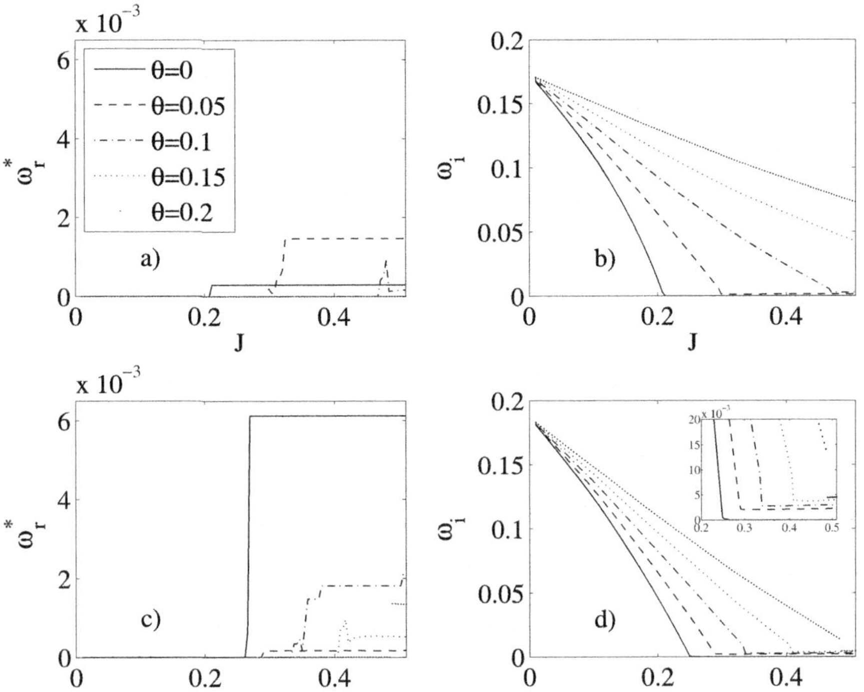

Figure 7.11: Results from linear stability analysis for $R = 1$, and different slope angles. Intrinsic frequency $\omega_r^* = \omega_r - a\alpha_r$ (left column) and temporal amplification rates (right column) with respect to the the Richardson number J, for $\alpha_r = 0.3$ ((a) and (b)) and $\alpha_r = 0.5$ ((c) and (d)). In (d) the area where the amplification rates drops to a constant value are zoomed in the additional square.

7.6.2 Inclined exchange flow with superimposed weak mean advection

The mean advection a in the temporal case acts as a Doppler shift in frequency, as it was shown by Ortiz and Loiseleux [2002], and it affects only the real part of ω in the temporal theory. Therefore, temporal instability will be fully described by considering the intrinsic frequency of the temporal mode, defined as the frequency of the wave seen by an observer moving with the local mean flow $\omega_r^* = \omega_r - a\alpha$ as a function of α. Moreover $\omega(\alpha) = -\bar{\omega}(-\bar{\alpha})$ (where $\bar{}$ denotes the complex conjugate) so that only positive wave numbers can be considered, without any loss of generality. All the results presented in the following are obtained for $a = 0.05$. Changes of the mean advection become important in the spatial stability analysis.

In figure 7.11 the intrinsic frequency (left column) and the temporal amplification rates (right column) are plotted versus the Richardson number J for $\alpha_r = 0.2$ ((a) and b)) and for $\alpha_r = 0.5$ ((c) and d)) for different values of the bottom slope. For a horizontal bottom ($\theta = 0$, continuous line) there is initially a strong decrease of the temporal amplification rate for increasing Richardson numbers. At a certain value of the Richardson number, the curve reaches a zero value. If the buoyant component containing the Richardson number is decomposed in one component parallel and one orthogonal to the inclined bottom, the

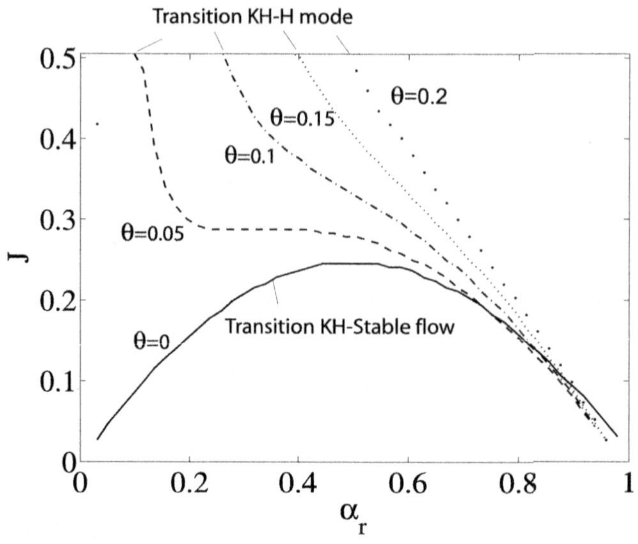

Figure 7.12: Transitions boundaries from KH to Holmboe modes for different slope angles, $R = 1$ and $a = 0.05$. The instability regions are increased for larger slope angles. For $\theta > 0$ the curve represent the transition from KH to Holmboe modes, while for $\theta = 0$ the curve represents the stability boundary as the amplification rates of the Homlboe mode are zero, thus indicating a neutral mode.

temporal amplification rates do not decrease to zero in the range of Richardson numbers studied: they reach a very small value and remain constant. The details are shown in the zoom in the figure 7.11 (d). The constant value reached is larger for larger values of the bottom slope. Due to this fact, that the amplification rates does not become zero if $\theta \neq 0$, it means that the point at which the curves stop to decrease and become constant, is not the transition from unstable to stable modes, but it represents the transition from KH to Holmboe modes, characterized with a non-zero phase speed (cf. figure 7.11 (a) and (c)). It is also worth noting that the transition occurs at higher Richardson numbers if the slope is larger. Moreover, the decreasing rate of the amplification rates is lower if the slope increases. Holmboe waves are neutral in the case of a zero bottom slope, as the amplification rates becomes zero ((cf. figure 7.11 (b) and (d)), continuous lines). For $\theta = 0$, the transition is found when the amplification rates becomes zero, while if $\theta > 0$, the transition from KH to Holmboe modes is found at the point where the amplification rates reaches the constant value and the phase speed becomes non-zero. Each of these locations, at a given wavenumber, corresponds to a critical Richardson number. The transition boundary is plotted in figure 7.12 for the different slopes. The results for $\theta = 0$ are in good agreement with previous results (continuous line). If the slope angle is non-zero, the unstable regions where KH modes exist are larger particularly for smaller wavenumbers. The regions on the right hand side of the boundary lines for $\theta > 0$ in figure 7.12 are unstable regions too, in which Holmboe modes live. These results suggest that the bottom slope has a similar influence on the stability of the flow as the change of the parameter R for the density profile. In fact, from a physical point of view, we expect that both a bottom slope and an increased ratio R have the effect of destabilizing the flow, as both terms are modifying the contribution of the buoyancy term (which gives a stabilizing

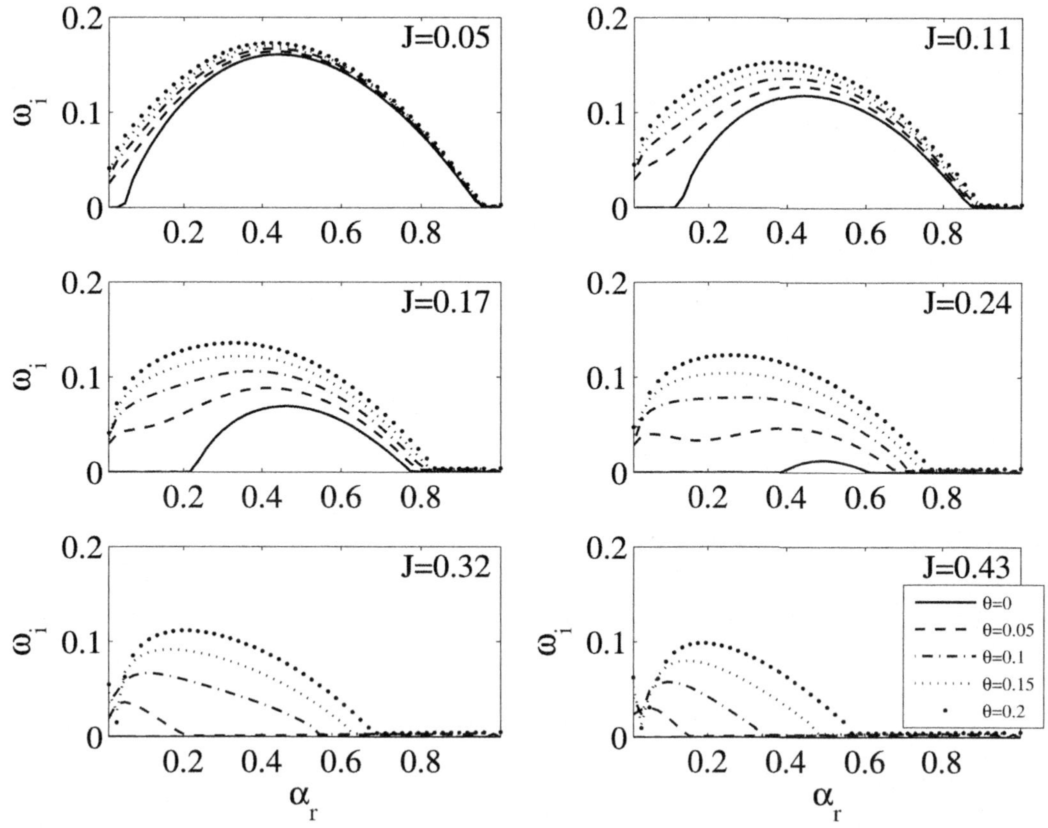

Figure 7.13: Transitions from KH to Holmboe modes for different slope angles, $R = 1$ and $a = 0.05$. The instability regions are increased for larger slope angles. While for $\theta > 0$, the curve represent the transition from KH to Holmboe modes, for $\theta = 0$ the curve represents the stability boundary.

contribute) in the stability equation (7.20). The difference is that with the bottom slope other terms are added into the governing stability equations, but their value remains small since the slope angles are small. If the amplification rates are plotted versus the wavenumber as illustrated in figure 7.13 it can be noted that, for a given Richardson number, there is a range of wavenumbers in which the amplification rates are particularly large. These are the regions of maximal amplification rates and correspond to the KH mode (characterized with a zero phase speed, see figure 7.11), whatever the value of the bottom slope. The difference between the curves for different slopes is very weak for low Richardson numbers; as it increases the difference increases much more for the horizontal case as compared to the inclined one. While for the horizontal bottom, the range of wavenumbers with non zero amplification rates is shifting to the middle point $\alpha_r = 0.5$, only the right boundary is shifting to the left for the inclined cases, thus indicating that for larger Richardson numbers, i.e. for more stable flows, the bottom slope makes the flow more unstable relative to long waves, while shorter waves becomes more and more indifferent. This is also shown in figure 7.14, in which the maximal amplification rates (a) and the correspondent wavenumber (b) are plotted versus the Richardson numbers.

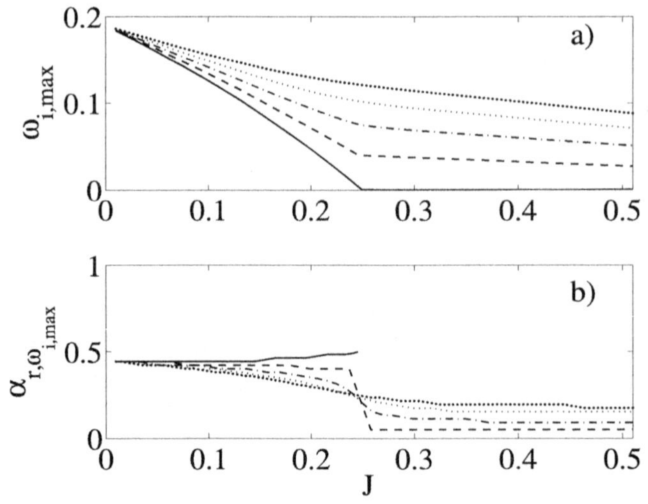

Figure 7.14: Maximal amplification rates (a) and correspondent wavenumber (b) for different Richardson numbers. The maximal amplification rates are found only for the KH modes. Holmboe´s amplification rates are at least one order of magnitude smaller.

For larger Richardson numbers, the maximal amplification rates are decreasing, reaching a zero value for the horizontal flow, and continuously decreasing for the inclined flow. The decreasing rate is lower for larger slope angles. It is worth noting that these maximal amplification rates correspond to KH modes, while Holmboe´s amplification rates are at least one order of magnitude smaller.

Also interesting to note is that while the correspondent value of the wavenumber, at which the most amplified mode develops, is slightly increasing for the horizontal flow reaching the well known value 0.5, at $J = 0.25$, the other curves present an inflection point at $J = 0.25$ and the values are decreasing for larger Richardson numbers. For $J < 0.25$, the value of the wavenumber correspondent to the most amplified mode decreases if the slope is larger, while for $J > 0.25$ larger slopes have larger wavenumbers correspondent to the highest amplification rates. At $J = 0.25$, for all slopes these value of the wavenumber collapse at the same value $\alpha_r = 0.25$. At this point also the curves of the amplification rates for $\theta > 0$ changes and their slope becomes smaller.

7.7 Summary and conclusions

A first step in understanding the stability of a two-layer stratified exchange flow down a slope was done by performing a linear stability analysis, through which a first insight in this method of analysis could be obtained. The modeled flow for this analysis is strongly idealized as many simplifications and assumptions have been met. It thus does not take into account the flow processes occurring at the interface. These are for example the unsteady barotropic forcing superposed on the baroclinic exchange flow, the fact that the flow is non-parallel and that is varying not only in the vertical but also in the streamwise direction (two-dimensional variability). The obtained experimental results presented in the previous chapters show that these effects are dominating the flow behavior and particularly the generation of instabilities at the interface. Comparisons to experimentally measured quantities are done only for validating the choice of certain definitions for the stability analysis, as for example the velocity profiles and other parameters characterizing the shear layer or density layer thickness. The results relative to the stability analysis which includes amplification rates, wavenumbers and frequencies are not compared to experimental data, since the flow defined for the stability analysis strongly differs from the high-turbulent, baroclinic/barotropic flow analyzed with the experimental investigations. Nevertheless, linear instability does correctly describe the onset and early evolution of infinitesimal perturbations and the results obtained can be used as starting point for a non-linear analysis.

Different numerical methods were tested, namely a pseudospectral method employing Chebyshev Polynomials and a finite difference method. Finally, a central, second-order finite difference method with an irregular grid was employed. The stability was studied from a temporal frame of reference. From the results it was concluded that for a given wavenumber, the bottom slope seems to have two effects on the stability of the flow: firstly, it increases the unstable regions as the stationary neutral stability curve is not the stability boundary but the boundary for the transition from KH modes to Holmboe modes. Secondly, it introduces in the unstable region the coexistence of two types of instabilities: KH instabilities characterized with higher temporal amplification rates and which decrease for increasing Richardson numbers, and Holmboe instabilities, which lives at higher Richardson numbers and which have a finite, non-zero phase speed. At smaller wavenumbers the regions in which KH modes exist are increased as compared to the horizontal case and the amplification rates are also found to be larger and to decrease slower with increasing Richardson numbers. The amplification rates of the Holmboe modes are found to be much smaller (one order of magnitude) as compared to those of the KH modes, and they are independent on changes of the wavenumber.

8 Concluding remarks

In this thesis, the interfacial wave activity at the interface between the upper and lower layer of a density-stratified exchange flow down a topographical obstruction has been studied. The effects of different experimental initial conditions, and so the effect of a pulsating barotropic component superimposed to the baroclinic flow, have been analyzed revealing that pulsating density currents are generated in the lower layer in order to balance the barotropic generated overflow between the two reservoirs. These surges-like flows form in their wakes high turbulent recirculation zones, thus causing significant entrainment and mixing between the two layers. The size of these structures has shown to be comparable to the total water depth. The experiments have all been run with Reynolds numbers in the order of $10^3 \div 10^4$, thus they are more representative for natural flows as the flow exchange between sea straits and between two reservoirs connected by a narrow pass in environmental engineering, in comparison with past studies. Also the investigated superimposed barotropic flow is of relevance for natural flows. For connections between two reservoirs of different densities, like the Burlington Ship Canal in Canada or the Lake Lucerne in Switzerland, the pulsating barotropic components are dictated by basin oscillations or seiches, and their period is triggered by the geometrical dimensions of the reservoirs and of the connecting channel. In this respect, the results of the present work are very representative with similar periods and can predict not only the size of the interfacial waves and the related entrainment and mixing at the interface for different Richardson and Reynolds numbers, but also the net flow rate oscillations and the related period, simply as a function of the geometrical conditions. Moreover, the results are also representative for the oceanographical or geophysical community: for infinitely large water bodies connected by a narrow pass or channel, like sea-straits, the pulsating barotropic component superposed to the baroclinic flow is dictated by metereological or astronomical tides so that again, the results of this study relative to the interfacial wave activity and the entrainment coefficients are also representative for such flows (see for example the studies of Sherwin and Turrell [2005]), even if the time scales in the model are faster than in nature.

Control methods in form of local enhanced bottom roughness in order to influence the generation and the developing of the large-scale structures at the interface, have shown to influence not only the developing mechanism of the pulsating density currents, but also their period of generation. Two different types of bottom roughness have been used during the experiments, which generated different types of velocity profiles. If the velocity profile is such that the bottom roughness causes a very strong velocity gradients near the bottom, the wakes behind the surges are not generated anymore. Nevertheless, entrainment and mixing are in the same order of magnitude as in the smooth case, due to the strong Reynolds stresses, causing a much rapid decay of turbulence and a more effective redistribution of turbulent kinetic energy in the three coordinate directions. If the velocity gradients in the lower layer are not so strong, the large-scale structures still generate, but they can not fully develop the wake region, so that both the size of the structures and the entrainment coefficients have shown to be reduced as compared to the smooth case. Also these results are representative for natural flows, not only in the geophysical context, but also for atmospheric flows, like windstorms down mountains or the air exchange and circulation in cities in a valley between night and day (see for example studies of Cermak [1993]). For example, winds in the atmosphere in which air cooled by contact with a cold ground flows down a hill are of importance to agriculture because they can influence damages due to frost or they can carry smoke or fog, thus affecting the visibility at ports.

However, the present work also contains some uncertainties. In the PIV measurements conducted in this thesis, the size of one pixel on the CCD chip was about half a millimeter (cf. section 3.3). This small ratio of particle size to pixel size (about 1/3 instead of the suggested 2/1) was inevitable in these experimental setup due to the large field of view required for the study. This inconvenience could be partly suppressed by defocusing the field of view with the lens mounted on the CCD camera, leaving to a ratio of 1 in the worst scenario. The small peak locking observed for all the experiments can be justified by the fact that the software DaVis interpreted sometimes a group of particles as one single particle, thus removing the inconvenient of small ratio. Of course, no structures smaller than roughly 1 cm x 1 cm could be resolved, but this was also not the scope of this work, since the attention was directed in understanding the large-scale governing processes occurring at the interface. Due to the various difficulties encountered during the calibration procedure for the PLIF system, the combined PIV/PLIF measurements could not deliver the Reynolds transport terms $\overline{u'c'}$, as the precision of the concentration measurements was not sufficient to determine accurate concentration fluctuations c'. However, the mean concentration fields could be used to describe the most important properties of the flow. For the experiments with local enhanced bottom roughness, a very rough estimation of the roughness coefficient has been made so that the values are rudimentary and no prediction of entrainment coefficients and size of the large-scale structures observed at the interface could be made as a function of a drag or roughness coefficient.

Another point which awakes a lot of interesting questions is the connection between the period of oscillation of the net flow rate and the period of generation of the large-scale surges at the interface: as pointed out in chapter 5, experimental results have demonstrated that these two periods are not exactly the same. Moreover, from experimental observations, it has been noted that this difference changes when enhanced bottom roughness is locally brought above the sill and when the water depth is changing. The variation of the generation period with the variation of the water depth or with increasing bottom roughness should be related with a general expression. For this aim, detailed PIV measurements with different water depths and purposely designed barotropic oscillations are required. Such studies are currently in progress at the Institute for Hydromechanics.

Also the effect of the sill slope and so of different acceleration components on the generation period could have an influence and could be further investigated. The effect of higher Reynolds numbers, in the order of $10^5 \div 10^6$ using higher water depths and buoyant accelerations should be further investigated. This surely will demand high speed cameras with a high temporal, but also spatial resolution. As these experiments will be very challenging from the technical point of view, an interesting compromise would be to perform numerical simulations using for example DNS or LES, also in order to verify the experimental data. In this view, the experimental data collected during the course of this work could deliver the data necessary to calibrate numerical codes, so that various numerical simulations could be run investigating the effect of high Reynolds numbers on the interfacial wave activity. These results could also furnish more information about an eventual Reynolds invariance.

A first step in understanding the stability of a two-layer stratified exchange flow down a slope was done by performing a linear stability analysis, through which a first insight in this method of analysis could be earned. However, careful attention should be paid for these results, which are useful for the overall understanding of the flow stability and which can correctly predict only the early evolution of infinitesimal perturbations. Moreover, the modeled flow for this analysis is strongly idealized as many different simplifications and assumptions have been met and as such does not take into account the full portrait of the flow processes occurring at the interface. These are for example the unsteady barotropic forcing superposed on the baroclinic exchange flow, the fact that the flow is non-parallel and that is varying not only in the vertical but also in the streamwise direction (two-dimensional variability). The obtained experimental results show that these effects are dominating the flow behavior and particularly the generation of instabilities at the interface. For these reasons, results relative to the stability analysis, such as amplification rates and wavenumbers, were not compared to experimental data. Comparisons to experimentally measured quantities are done only for validating the choice of certain definitions for the stability analysis, as for example the velocity profiles and other parameters characterizing the shear layer or density layer thickness. Nevertheless, linear instability does correctly describe the onset and early evolution of infinitesimal perturbations and the results can be used as starting point for a non-linear analysis.

From the results of the linear analysis it is concluded that for a given wavenumber, the bottom slope has two effects on the stability of the flow: first, it increases the unstable regions as the stationary neutral stability curve is not the stability boundary anymore. Second, it introduces in the unstable region the coexistence of two types of instabilities: KH instabilities characterized with higher temporal amplification rates and which decrease for increasing Richardson numbers, and Holmboe instabilities, which lives at higher Richardson numbers and which have a finite, non-zero phase speed. Further analysis is required to study the additional effect of the viscosity and of the change of the ratio of the shear layer thickness to the density layer thickness. Also the effect of bottom roughness on the stability of the flow, using a different definition for the velocity profile, is an issue that still has to be addressed.

Bibliography

P. W. Aeschbach. Investigation of the flow behind a two-dimensional model with blunt trailing edge and fitted with splitter plate. *J. Fluid Mech.*, 21:241–255, 1996.

Y. D. Afanasyev and W. R. Peltier. On breaking internal waves over the sill in knight inlet. *Proc. Royal Society*, 457(2016):2799–2825, 2001.

G.H. et al Alavian, V. Jirka. Density currents entering lakes and reservoirs. *J. Hydraulic Engineering, ASCE*, 118(11):1464–1489, 2002.

G. Anati, D. A. Assaf and R.O.R.Y. Thompson. Laboratory models of sea straits. *J. Fluid Mech.*, 81:341–351, 1977.

M. Arita and G.H. Jirka. Two layer model of saline wedge. i: Entrainment and interfacial friction. *J. Hydraulic Engineering*, 113:1229, 1987.

L. Armi. The hydraulics of two flowing layers with different densities. *J. Fluid Mech.*, 163:27–58, 1986.

L. Armi and D. M. Farmer. Maximal two-layer exchange through a contraction with barotropic net flow. *J. Fluid Mech.*, 164:27–58, 1986.

G. Assaf and A. Hecht. Sea straits: a dynamical model. *Deep-Sea Research*, 21:947–958, 1974.

J. Atkinson. Interfacial mixing in stratified flows. *J. Hydr. Res.*, 26(1):27–31, 1988.

W. D. Baines. Entrainment by a plume or a jet at a density interface. *J. Fluid Mech.*, 68:309–320, 1975.

W. D. Baines. *Topographic effects in stratified flows.* Cambridge University Press, Cambridge, 1989.

W. D. Baines. *Topographic effects in stratified flows.* Cambridge University Press, Cambridge, 1995.

N. Baker and P.F. Linden. Physical modeling of air flows: a new design tool. *Atr. Build. Arch. Eng.*, pages 13–22, 1991.

A. Bayliss, A. Class and B.J. Matkowsky. Roundoff error in computing derivatives using the chebyshev differentiation matrix. *J. Comput. Phys.*, 116(2):380–383, 1995. ISSN 0021-9991. doi: http://dx.doi.org/10.1006/jcph.1995.1036.

T.B. Benjamin. Gravity currents and related phenomena. *J. Fluid Mech.*, 31:209, 1968.

J. P. Boyd. *Lecture Notes in Engineering 49: Chebyshev and Fourier Spectral Methods.* Springer Verlag, Heidelberg, Germany, 1989.

A.G. Brown, W.G. Wilson and K.R. Solvason. Heat and miosture flows through openings by convection. *J. Am. Soc. Heat. Vent. Air Cond. Eng.*, 5:49–54, 1963.

W.G. Brown and K.R. Solvason. Natural convection heat transfer through rectangular openings in partitions-i. *Int. J. Heat Mass Transf*, 5:859–868, 1962a.

W.G. Brown and K.R. Solvason. Natural convection heat transfer through rectangular openings in partitions-ii. *Int. J. Heat Mass Transf*, 5:869–878, 1962b.

C. Canuto, M.Y. Hussaini, A. Quarteroni, and T.A. Zang. *Spectral Methods in Fluid Dynamics.* Springer Verlag, NY, 1988.

J.E. Cermak. *Wind climate in cities.* Kluver Acad. Pub., ISBN 0792332024, 1993.

J. Cherman, F.S. Imberger and G.M. Corcos. Turbulence and mixing in stably stratified waters. *Ann. Rev. Fluid Mech*, 10:267–288, 1978.

G. C. Christodoulou. Interfacial mixing in stratified flows. *J. Hydraulic Research*, 24: 77–92, 1986.

G. C. Christodoulou. Dilution of dense effluents on a sloping bottom. *J. of Hydraulic Research*, 29(3):329–339, 1991.

A. J. Cotel and R. E. Breidenthal. Persistence effects in stratified entrainment. *Appl. Sci. Res.*, 57:349–366, 1997.

E.A. Cowen and S.G. Monismith. A hybrid digital particle tracking velocimetry technique. *Exp. in Fluids*, 22:199–211, 1997.

P. F. Crapper and P. F. Linden. The structure of turbulent density interfaces. *J. Fluid Mech.*, 65:45–63, 1974.

T.L. Criminale, W.O. Jackson and R.D. Joslin. *Theory and computation of hydrodynamic Instability.* Cambridge University Press, Cambridge, 2000.

P.A. Cuthbertson, A.J.S. Davies and M.J. et al Coates. A modeling study of transient, buoyancy-driven exchange flow over a descending barrier. *Environ. Fluid Mech.*, 4(2): 127–155, 2004.

J. Dallimore, C.J. Imberger and T. Ishikawa. Entrainment and turbulence in saline underflow in lake ogawara. *J. Hydraulic Engineering, ASCE*, 127(11):937–948, 2001.

S.B. Dalziel and G.F. Lane-Serff. Linear viscous stability theory for stably stratified shear flow: A review. *Buil. Env.*, 5(2):121–135, 1991.

P. J. W. Daviero, G. J. Roberts and K. Maile. Refractive index matching in large-scale stratified experiments. *Exp. in Fluids*, 31:119–126, 2001.

G.M.J. Davies. *Buoyancy driven flow through openings*. Ph. d. thesis, Cambridge University, UK, 1993.

R. G. Dean and R. A. Dalrymple. *Water wave mechanics for engineers and scientists*. Englewood Cliffs:Prentice-Hall, XI-353, 1984.

P.G. Drazin and W.H. Reid. *Hydrodynamic stability*. Cambridge University Press, US, 1981.

T.H. Ellison and J.S. Turner. Turbulent entrainment in stratified flows. *J. Fluid Mech.*, 6:423– 448, 1959.

M. Epstein. Buoyancy driven exchange flow through openings in horizontal partitions. *Intl. Conf. Cloud Vapor Modelling*, pages –, 1988.

D. M. Farmer and L. Armi. Maximal two-layer exchange over a sill and trough a combination of a sill and a contraction with barotropic flow. *J. Fluid Mech.*, 164:53–76, 1986.

D. M. Farmer and L. Armi. The establishment of stratified flows over topography by small scale entrainment and mixing. *Proc. Roy. Soc. London*, 455:3221–3258, 1999.

R. L. Fernandez and J. Imberger. Bed roughness induced entrainment in a high richardson number underflow. *J. Hydr. Eng.*, 26:27–31, 2007.

H. J. S. Fernando. Turbulent mixing in stratified fluids. *Ann. Rev. Fluid Mech.*, 23: 455–493, 1991.

J.H. Ferziger and M. Peric. *Computational methods for fluid mechanics*. Springer, Berlin Heidelberg, 1996.

D. S. Finnigan, T. D. Luther and R. Lukas. Observations of enhanced diapycnal mixing near the hawaiian ridge. *J. Phys. Oceanogr.*, 32(11):2988–3002, 2002.

K. B. Finnigan, T. D. Winters and G. N. Ivey. Response characteristics of a buoyancy-driven sea. *J. Phys. Oceanogr.*, 31:2721–2736, 2001.

T.D. Finnigan and G.N. Ivey. Convectively driven exchange flow in a stratified sill-enclosed basin. *J. Fluid Mech.*, 418:313–338, 2000.

H.B. etal Fischer. *Mixing in Inland and Coastal Waters*. Academic Press, New York, 1979.

B. Fornberg. Calculation of weights in finite difference formulas. *SIAM Review, Society for Industrial and Applied Mathematics*, 40(3):685–691, 1998.

B. Fornberg. Generation of finite difference formulas on arbitrarily spaced grids. *Mathem. of Comp.*, 51(184):699–207, 1988.

H.R.S. Fouli. *An experimental study of interfacial waves and instabilities in exchange flows over a smooth sill*. Dep. of civil and environmental engineering, University of Alberta, Edmonton, Canada, 2006.

H.R.S. Fouli and D.Z. Zhu. Interfacial waves in exchange flows over a sill. *Environmental Hydraulics and Sustainable Water Management - Lee and Lam*, 0415365465, 2005.

C. Francis, P. Oppenheimer and D. Stevenson. Endogenous growth of persistently active volcanoes. *Nature*, 366:554–557, 1993.

D.C. Fugate and R.J. Chant. Near-bottom shear stresses in a small, highly stratified estuary. *J. of Geophys. Res. Oceans*, 110(C3):3022, 2005.

K.S. Gage. Linear viscous stability theory for stably stratified shear flow: A review. *Boundary-Layer Meteorology*, 5(1-2):3–17, 1973.

M. H. Garcia. Hidrodinamica ambiental. *Publicaciones Universidad National del Litoral*, page 197, 1996.

M. Gaster. A note on the relation between temporally-increasing and spatially-increasing disturbances in hydrodynamic stability. *J. Fluid Mech.*, 14(2):222–224, 1962.

A.Yu. Gelfgat and E. Kit. Spatial versus temporal instabilities in a mixing layer. *J. Fluid Mech.*, 552:189–227, 2006.

C. Gerdes, F. Garrett and D. Farmer. On internal hydraulics with entrainment. *Journal of Physical Oceanography*, 32(3):1106–1111, 2002.

M. Ghisalberti and H.M. Nepf. Canopies. *Water Resources. Res.*, 40(7):197, 2004.

S. Goldstein. On the stability of superposed streams of fluids of different densities. *Proc. of R. Soc. of London*, 132(A):524–548, 1931.

L. Gu and G. A. Lawrence. Analytical solution for maximal frictional two-layer exchange flow. *J. Fluid mech.*, 543:1–17, 2005.

S. Haigh and G. A. Lawrence. Symmetric and non-symmetric holmboe instabilities in an inviscid flow. *Phys. of Fluids*, 6:1459–1468, 1999.

P.F. Hamblin and G. A. Lawrence. Exchange flows between hamilton harbour and lake ontario. *Proc. of 1990 Annual Conf. of Canadian Society for Civil Eng.*, 5:140–148, 1990.

B. Hansen and S. Osterhus. North atlantic-nordic seas exchanges. *Progress in Oceanogr.*, 45(2):109–208, 2000.

P. Hazel. Numerical studies of the stability of inviscid stratified flows. *J. Fluid mech.*, 6 (–):39–61, 1972.

J. Hebbert, B. Imberger and J. Loh, I. Patterson. Collie river underflow into the wellington reservoir. *J. Hydr. Div.*, 105(5):533–545, 1979.

K.A. Helfrich. Time-dependent two-layer exchange flows. *J. Phys. Oceanogr.*, 25:359–373, 1995.

A. Hogg and G. Ivey. The kelvin-helmholtz to holmboe instability transition in stratified exchange flows. *J. Fluid Mech.*, 477:339–362, 2003.

J. Holmboe. On the behavior of symmetric waves in stratified shear layers. *Geophysiske Publikasjoner*, 6:39– 61, 1962.

P. Huerre and P.A. Monkewitz. Local and global instabilities in spatially developing flows. *Ann. Rev. Fluid Mech.*, 22(22):473, 1990.

H.E. Huppert and M.A. Hallworth. Bi-directional flows in constrained systems. *J. Fluid Mech.*, page in press, 2006.

G.H. Jirka and M. Arita. Density currents or density wedges: boundary layer influence and control methods. *J. Fluid Mech.*, 177:186–206, 1987.

R.H. Kase and A. Oschlies. Flow through denmark strait. *Journal of Geophysical Research - Oceans*, 105(C12):28527–28546, 2000.

D.E. Keil. *Buoyancy driven counterflow and interfacial mixing.* Ph. d. thesis, Cambridge University, UK, 1991.

T. Kessel and C. Kranenburg. Gravity current of fluid mud on sloping bed. *J. Hydr. Eng.*, 122(12):710–717, 1996.

M.M. Koochesfahani and P.E. Dimotakis. Laser induced fluorescence measurements of mixed fluid concentration in a liquid plane shear layer. *AIAA J.*, 23(11):1700–1707, 1985.

C.G. Koop and F.K. Browand. Instability and turbulence in a stratified fluid with shear. *J. Fluid Mech.*, 24:67–113, 1976.

Y. A. Kuznetsov. *Elements of applied bifurcation theory.* Springer Verlag, NY, 1998.

F.K. Lawrence, G. A. Browand and L. Redekopp. The stability of a sheared density interface. *Phys. Fluids*, 3:2360–2370, 1991.

G. A. Lawrence. The hydraulics of steady two-layer flow over a fixed obstacle. *J. Fluid Mech.*, 254:605–633, 1993.

G.T. Lien, R.C. D Asaro and G.T. Dairiki. Lagrangian frequency spectra of vertical velocity and vorticity in high reynolds number oceanic turbulence. *J. Fluid Mech.*, 362:177–198, 1998.

G.F. Linden, P.F. Lane-Serff and D.A. Smeed. Emptying filling spaces: the fluid mechanics of natural ventilation. *J. Fluid Mech*, 212:300–335, 1990.

P.F. Linden. The fluid mechanics of natural ventilation. *Ann. Rev. Fluid Mech*, 31: 201–238, 1999.

K. Lofquist. Flow and stress near an interface between stratified fluids. *Phys. Fluids*, 3 (2):158–169, 1960.

R.R. Long. Some aspects of the flow of stratified fluids. ii. experiments with a two-fluid system. *Tellus*, 6:97, 1954.

S.A. Maslowe. *Shear flow instabilities and transition in hydrodynamic instabilities and the transition to turbulence.* Springer Verlag, Berlin, 1985.

J.W. Miles. On the stability of heterogeneous shear flows. *J. Fluid Mech.*, 10:496, 1961.

D.Z. Morin, V.M. Zhu and M.R. Loewen. Supercritical flow down a sill. *J. Hydraulic Engineering*, 130(6):11 pages, 2004.

D. Myrhaug and O.H. Slaattelid. Bottom shear stresses and velocity profiles in stratified planetary boundary layer flow from similarity theory. *J. of MArine Systems*, 14(1-2): 167–180, 1998.

D.Z. Negretti, M.E. Zhu and G.H. Jirka. Barotropicly induced interfacial waves in exchange stratified flows over a smooth sill. *J. Fluid Mech.*, 2007.

E. Oguz, T. Ozsoy and M. et al Latif. Modeling of hydraulically exchange flow in the bosphorus strait. *J. Phys. Oceanogr.*, 20:945–965, 1990.

J.M. Ortiz, S. Chomaz and T. Loiseleux. Spatial holmboe instability. *Phys. Fluids*, 14 (8):2585–2596, 2002.

P.R. Owen. Buildings in the wind. *Q.J.R.Met.Soc.*, 97:396–413, 1971.

M. Parker, G. Garcia and W. Fukushima, M. Yu. Experiments on turbidity currents over an erodable bed. *J. Hydraulic Engineering*, 25:123–147, 1987.

P.F. Patterson, J.C. Hamblin and J. Imberger. Classification and dynamic simulation of the vertical density structure of lakes. *Limnol. Oceanogr.*, 29(4):845–61, 1984.

G. Pawlak and L. Armi. Mixing and entrainment in developing stratified currents. *J. Fluid. Mech.*, 424:1–37, 2000.

G. Pawlak and L. Armi. Vortex dynamics in a spatially accelerating shear layer. *J. Fluid. Mech.*, 376:1–35, 1998.

C. Pellacani. Stability analysis of stratified sheared flows as an initial-value proble. *Nuovo Cimento B Serie, Provided by the Smithsonian/NASA Astrophysics Data System*, 33 (A):569–578, 1976.

W.R. Peltier and C.P. Caulfield. Mixing efficiency in stratified shear flows. *Annual Rev. Fluid Mech.*, 35:135–167, 2003.

W.H. Perry, A.E. Schofield and P.N. Joubert. Rough wall turbulent boundary layers. *J. Fluid Mech.*, 37:383–413, 1969.

L.J. Pratt. Rotating shocks in a separated laboratory channel flow. *J. of Phys. Ocean.*, 17:483–491, 1987.

M.A. Reidenbach and A. Genin. Boundary layer turbulence and flow structure over a fringing coral reef. *J. Limn. and Ocean.*, 51(5):1956–1968, 2006.

A. Sannino, G. Bargagli and V. Artale. Numerical modeling of the mean exchange through the strait of gibraltar. *Journal of Geophysical Research*, 107(C8):3094, 2002.

J.B. Schijf and J.C. Schoenfeld. Theoretical considerations on the motion of salt and fresh water. *Proc. Minnesota Int. Hydr. Con-ference, Joint Meeting IAHR and Hydr. Div., ASCE*, 0:321–330, 1953.

J.F. Scinocca and W.R. Peltier. Pulsating downslope windstorms. *J. Atmos. Sci.*, 46(-): 2885–914, 1989.

T.B. Sherwin and W.R. Turrell. Mixing and advection of a cold water cascade over the wyville thomson ridge. *Deep Sea Res.*, 1(52):1392–1413, 2005.

E.D. Skyllingstad and H.W. Wijesekera. Large-eddy simulation of flow over two-dimensional obstacles: high drag states and mixing. *J. of Phys. Ocean.*, 34(1):94–112, 2004.

G.P. Smyth, W.D. Klaassen and W.R. Peltier. Finite amplitude holmboe waves. *Geophys. Astrophys. Fluid Dyn.*, 43:181–222, 1988.

W.D. Smyth and W.R. Peltier. The transition between kelvin-helmholtz und holmboe instability; an investigation of the over reflection hypothesis. *J. Atmospheric Sciences*, 46:3698–3720, 1989.

W.D. Smyth and W.R. Peltier. Instability and transition in finite-amplitude kelvin-helmholtz and holmboe waves. *J. Fluid Mech.*, 228:387–415, 1991.

J.H. Stefan and T.R. Johnson. Negatively buoyant flow in a diverging channel iii: onset of underflow. *J. Hydr. Eng.*, 115:423–456, 1989.

A. Stigebrandt. On the effect of barotropic current fluctuations on the two-layer transport capacity of a constriction. *J. Phys. Oceanogr.*, 7:118–122, 1976.

H. Stommel and H. G. Farmer. Control of salinity in an estuary by a transition. *J. Mar Res.*, 12:13–20, 1953.

E.J. Strang and H.J.S. Fernando. Vertical mixing and transports through a stratified shear layer. *Journal of Physical Oceanography*, 31(8):2026–2048, 2001.

G. D. Sullivan and E. J. List. On mixing and transport at a sheared density interface. *J. Fluid Mech.*, 273:213–239, 1994.

G. Taylor. Effect if variation in density on the stability of superposed streams of fluids. *Proc. of R. Soc. of London*, 132(A):499–523, 1931.

S. Thorpe. Experiments on instability and turbulence in a stratified shear flow. *J. Fluid. Mech.*, 61:731–751, 1973.

S. Thorpe. Laboratory observation of secondary structures in kelvin-helmholtz billows and consequences for ocean mixing. *Geophys. Astrophys. Fluid Dyn.*, 35:175–199, 1985.

S. A. Thorpe. A laboratory study of stratified accelerating shear flow over a rough boundary. *J. Fluid. Mech.*, 138:185–196, 1983.

S.A. Thorpe. Transitional phenomena and the development of turbulence in stratified fluids: A review. *J. Geophysical Research*, 92:5231–5248, 1987.

C.D. Troy and J.R. Koseff. The generation and quantitative visualization of breaking internal waves. *Exp. in Fluids*, 38:549–562, 2005.

J.S. Turner. The influence of molecular diffusivity on turbulent entrainment across a density interface. *J. of Fluid Mech.*, 33:639–656, 1968.

P.J. Wallace. Volcanic so2 emissions and the abundance and distribution of exsolved gas in magma bodies. *J. Volcan. and Geotherm. Res.*, 510:145–168, 2003.

M.V. Welander. Flow around circular cylinder-kaleidoscope of challenging fluid phenomena. *Proc. Symp. Fully separated Flows*, 1:102, 1984.

G. B. Whitham. Variational methods and applications to water waves. *Proc. Ray. Soc. Lon.*, 229(-):6–25, 1967.

D.J. Wilson and D.E. Keil. Gravity-driven counterflow through an open door in a sealed room. *Buil. Env.*, 25:379–388, 1990.

A. Wuest and D.M. Farmer. In: Encyclopædia of science and technology, 9th edition. *McGraw-Hill*, 0, 2001.

D.M. Wuest, A. Imboden and M. Schurter. Origin and size of hypolimnic mixing in urnersee, the southern basin of vierwaldstättersee (lake lucerne), schweiz. *Schweiz. Z. Hydrol.*, 50:40–70, 1988.

G.A. Zaremba, L.J. Lawrence and R. Pieters. Frictional two-layer exchange flow. *J. Fluid. Mech.*, 474:339–354, 2003.

D.Z. Zhu. Control curves for two-layer flows. *J. Hydraulic Engineering*, 128(1):113–116, 2002.

D.Z. Zhu and G.A. Lawrence. Hydraulics of exchange flows. *J. Hydraulic Engineering*, 126(12):921–928, 2000.

D.Z. Zhu and G.A. Lawrence. Holmboe's instability in exchange flows. *J. Fluid. Mech.*, 429:391–409, 2001.

D.Z. Zhu and G.A. Lawrence. Non-hydrostatic effects in layered shallow water flows. *J. Fluid. Mech.*, 355:1–16, 1998.

Curriculum Vitæ

General

December 5, 1978	Born in Brunico, Italy
Permanent address	Via J. Ferrari 3H, 39031, Brunico (BZ), Italy
E-mail address	elettina@yahoo.it

Education

July 5, 1997	Secondary school diploma in Economics at the Liceo A. Cantore, Brunico, Italy
July 25, 2003	Graduation in Environmental Engineering at the University of Trento, Italy Title of thesis: Analysis of the wake behind a circular cylinder in shallow water flows
October 1, 2003–current	Research and teaching assistant at the University of Karlsruhe, Germany

Fields of study

Major field	Fluid Mechanics
Speciality area	Stratified exchange flows
Investigation methods	Experiments (PIV,PLIF) and linear stability analysis

Studies in Fluid Mechanics	Prof. GH Jirka
Studies in Turbulence	Prof. GH Jirka and Prof. W Rodi
Studies in Stratified Flows	Prof. DZ Zhu, Prof. GH Jirka
Studies in rotating effects and non non-Newtonian Fluids	Prof. R Bohning
Studies in Chaos theory and stability analysis	Prof. SA Socolofsky, Prof. A Class
Studies in numerical methods	Prof. W Rodi, Prof. SA Socolofsky
Studies in mathematics	Prof. J Fenton, Prof. P Ehrhardt

Publications

Journals

Negretti ME, Socolofsky SA, Rummel AC and Jirka GH (2005) *Stabilization of cylinder wakes in shallow water flows by means of roughness elements: an experimental study*, EXP. IN FLUIDS, **38**:403–413.

Negretti ME, Vignoli G, Tubino M and Brocchini M (2006) *On shallow wakes: an analytical study*, J. FLUID MECH., **567**:457–475.

Negretti ME, Zhu DZ, and Jirka GH (2007) *Barotropically induced interfacial waves in stratified two-layer flows over a sill*, J. FLUID MECH.,(in press).

Negretti ME, Zhu DZ, and Jirka GH (2007) *The effect of bottom roughness in two-layer flows down a slope*, submitted to J. PHYS. OCEAN.

Conferences

Negretti ME, Vignoli G (2004) *Scia Turbolenta su acqua bassa*, XXI CONVEGNO DI IDRAULICA E COSTRUZIONI IDRAULICHE, Trento, Italy.

Negretti ME, Zhu DZ., Weitbrecht V, Jirka GH (2005) *Experiments on stratified exchange flows past a submerged sill*, PROCEEDINGS OF 30TH CONGRESS OF THE INTERNATIONAL ASSOCIATION OF HYDRAULIC ENGINNERING (IAHR), Coex Seoul, Corea.

Negretti ME, Jirka GH, Zhu DZ (2006) *Combined PIV/PLiF measurements in stratified exchange flows over a submerged sill*, 6th INT. SYMP. ON STRATIFIED FLOWS, Perth, Australia.

Negretti ME, Socolofsky SA, and Jirka GH (2007) *Linear stability analysis of spatially accelerating stratified shear flows*, PROCEEDINGS OF 32TH CONGRESS OF THE INTERNATIONAL ASSOCIATION OF HYDRAULIC ENGINNERING (IAHR), Venice, Italy.

Reports

Negretti ME and Jirka GH (2006) *Hydrodynamic Instabilities and mixing and entrainment in two-layer stratified exchange flows over a submerged sill*, TECHNICAL REPORT (DFG REPORT NR. 831), Project Nr. Ji18/12-1, 16 pages.

Printed in Dunstable, United Kingdom

91343658R00087